THE BIG BOOK OF

THE EARTH

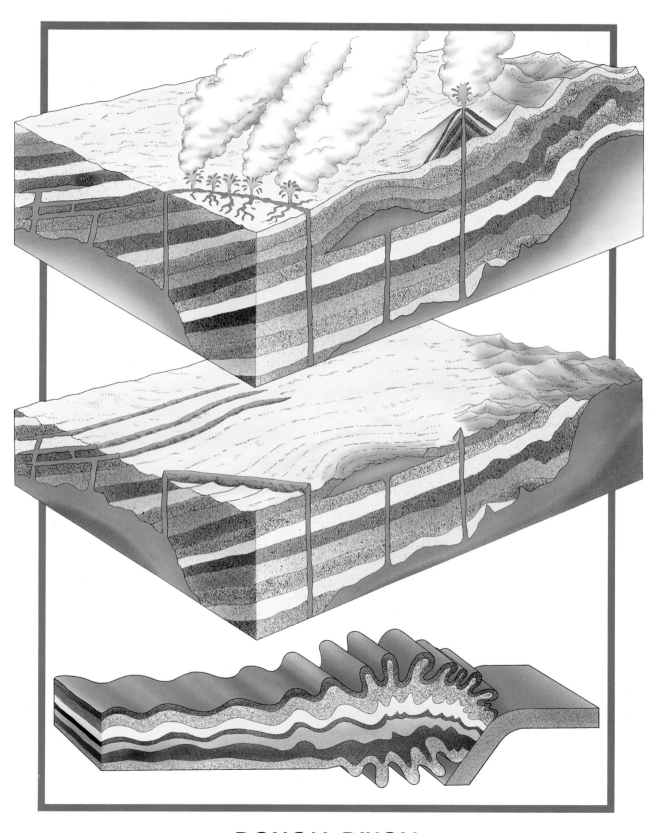

DOUGAL DIXON

SMITHMARK

This edition published in 1991 by SMITHMARK Publishers Inc.,
112 Madison Avenue, New York, NY 10016
By arrangement with Reed International Books
Michelin House, 81 Fulham Road, London SW3 6RB

ISBN 0-8317-0855-7

Printed in Italy

SMITHMARK Books are available for bulk purchase
for sales promotion and premium use.
For details write or telephone the Manager of Special Sales,
SMITHMARK Publishers Inc., 112 Madison Avenue,
New York, NY 10016. (212) 532-6600

CONTENTS

THE GROWING PLANET

Earth Science 8
How the Earth Formed 10
Inside the Earth 12
Earth's Magnetic Field 14
Movement of the Crust 16
Continents 18
Volcanoes 20
Earthquakes 22

THE SUBSTANCE OF THE EARTH

Minerals 24
Igneous Rocks 26
Sedimentary Rocks 28
Metamorphic Rocks 30
Folds and Faults 32
Mountains 34

THE PLANET'S HISTORY

Earth's Time Scale 36
Reading Earth's Story 38
Fossils and Evolution 40

THE CHANGING SURFACE

The Water Cycle 42
Rivers 44
Glaciers 46
The Seas 48
Seashores 50
Weathering and Erosion 52

OUR ENVIRONMENTS

Energy from the Earth 54
Alternative Energy 56
The Atmosphere 58
Weather and Seasons 60
Rainforests 62
Grasslands 64
Deserts 66
Temperate Zones 68
Coniferous Forests 70
Polar Regions 72
This Fragile Earth 74

Index 76
Acknowledgments 77

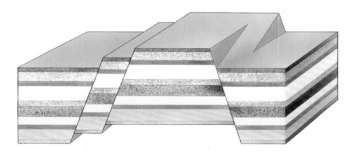

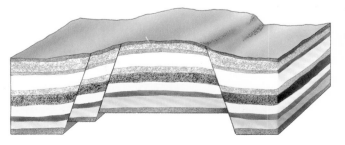

EARTH SCIENCE

Above: Ever since the first signs of civilization, people have been looking at the Earth and the heavens to try and find out how they formed and what they were made from.

The study of the Earth is fascinating because it covers the whole range of human knowledge. We need chemistry to discover what the Earth is made of. We need biology to interpret fossils and see how the life of our planet has changed. We need physics to work out what forces have changed the rocks since they were first formed so long ago.

If we build a house, we must know what kind of ground to put it on so that it does not fall down. Also we must know where to find the right kind of rocks for building stone, and the right kind of clay for bricks. If we grow crops, it is important to know if the soil is good enough, and if there is enough water flowing into the area. It is easy to see that civilization as we know it would be impossible without a proper understanding of the workings of the Earth.

EARLY SCIENTISTS

All this knowledge has come to us slowly. It was Leonardo da Vinci, in the fifteenth century, who first

realized that landscapes were shaped by rivers, and that fossils were the remains of ancient life. In the next century, the Italian scientist, Galileo, looked through his telescope and began to understand the movement of the Earth around the Sun. Other scientists also began to make discoveries, and soon all these studies were joined together into the science of geology (from the Greek *gaia*, the goddess of the Earth, and *logy* meaning study).

It was as geologists that Victorian scientists went out in their top hats, carrying hammers, to discover the basic ideas about how the Earth was formed. Then, in the middle of the twentieth century, satellites and spacecraft began to look at the Earth from a completely different viewpoint. The depths of the oceans began to be explored with machines that did not exist before. Gradually, the old science of geology was combined with oceanography, astronomy, and meteorology to give us the new and fascinating subject of Earth Science.

HOW THE EARTH FORMED

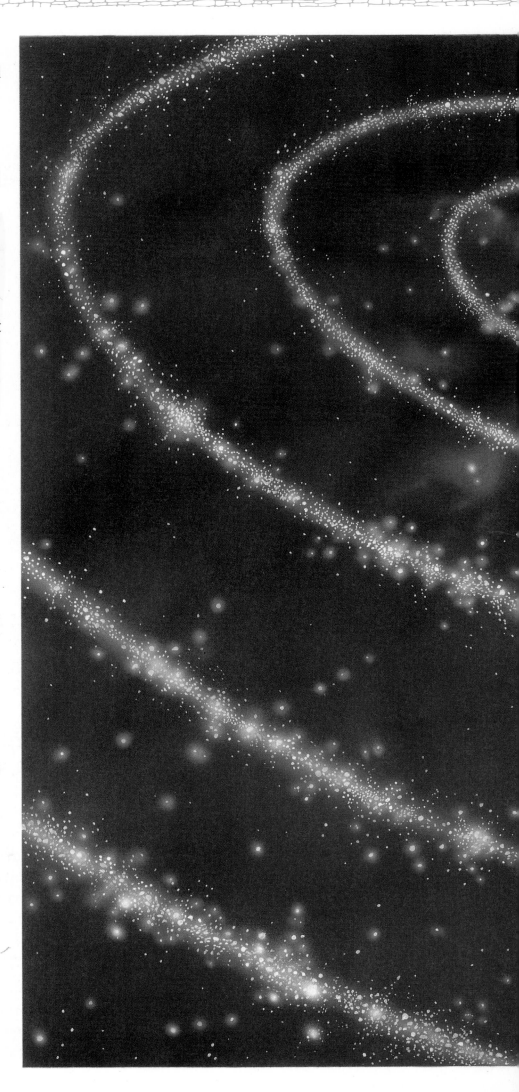

The Earth is one of a system of nine planets that revolve around the Sun. Most of the planets have moons that revolve around them. In addition, there are comets, asteroids and particles of dust. The whole arrangement is what we call the Solar System.

The Solar System was probably born as a cloud of dust floating in space. Gradually the cloud began to draw together, and as it did so it began to spin. This spinning spread it out into a disk. Most of the dust clustered at the center, and the energy produced by the colliding particles began to heat the cloud. The Sun was forming here.

BIRTH OF THE EARTH

Across the rest of the spinning disk, lumps of dust began to gather in eddies, and the planets began to form. There was heat produced here as well, but not as much as in the great central mass that was to be the Sun. As the Earth became solid, the heavier particles sank toward the center, and the lighter ones remained on the outside. The heat finally melted all the components into one layered ball.

As the surface cooled, much of the heat from the inside escaped through volcanoes. These eruptions brought steam to the surface and, when the surface was cool enough, the steam condensed as rain. The water eventually filled the hollows, and the first seas and oceans were formed. The lightest sorts of gas remained on the outside as the first atmosphere.

The Earth was born.

Above: Solid matter left over from the formation of the Solar System sometimes falls to Earth. As the pieces slice through the atmosphere, they burn up and glow, leaving a trail behind them. The result is what scientists call a meteorite. This big one was photographed in Wyoming in 1972.

INSIDE THE EARTH

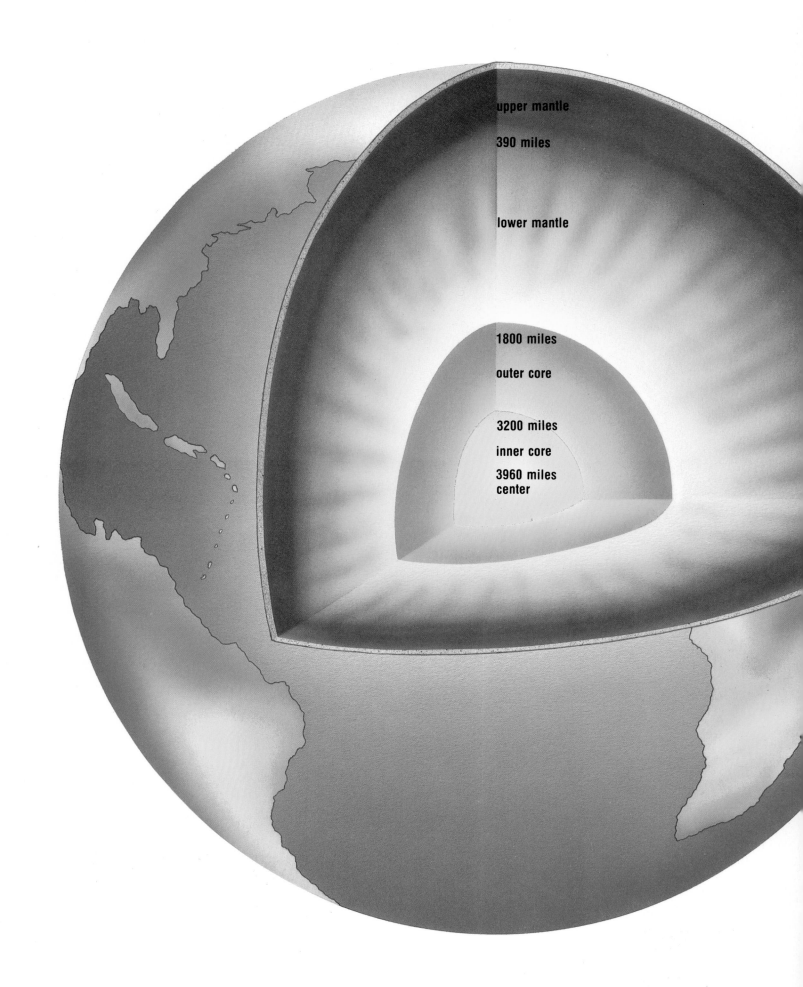

upper mantle

390 miles

lower mantle

1800 miles

outer core

3200 miles

inner core

3960 miles
center

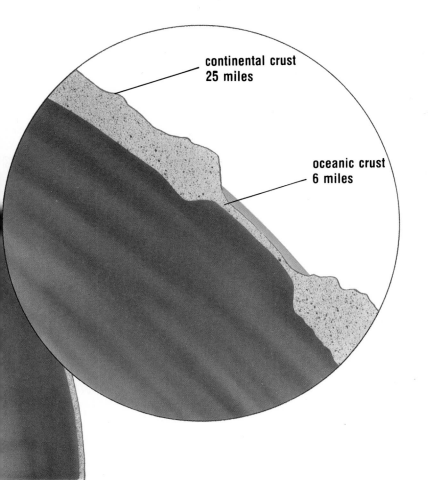

continental crust
25 miles

oceanic crust
6 miles

Above: The Earth consists of concentric layers, from dense iron at the core, to lighter and lighter stony materials toward the crust, with the lightest — gaseous matter — around the outside as an atmosphere.

Above: The roundness of the Earth is obvious to space travelers. Even in the low orbits of 125 miles or so used by the Space Shuttle, the horizon shows as a distinct curve.

We can best imagine the Earth as a giant jawbreaker. Like that candy, the Earth is made up of different layers. If we could lick away the top layer, we would find the one below to be made of different material, and the layers would change as we licked our way toward the middle.

Scientists do not know what these layers are made of – nobody has seen them – but we can make some fairly good guesses.

THE CENTER OF THE EARTH

At the very center of the Earth lies the core. This is made of very dense material, probably iron. It is extremely hot and would normally be liquid, but it is pressed so hard by the incredible weight of the layers around it that it may only be partially liquid. Scientists think that the inner core is solid, while liquid core material lies around it.

Around the outside of this core is the mantle, a lighter stony layer. It makes up most of the Earth and reaches almost to the surface. This mantle seems to be quite solid, but it may have a squishy layer near the top.

A THIN RIND

The outermost layer is the crust – the only part that we can see. This is made of even lighter material. There are two kinds of crust. The first – the oceanic crust – only lies at the bottom of the oceans. The second – the continental crust – is made of the lightest material of all. It lies embedded in the oceanic crust, on top of everything else.

EARTH'S MAGNETIC FIELD

Left: Highly charged particles blasted out of the Sun are caught by the Earth's magnetic field and pulled down to the Poles. There they react with the atmosphere to produce the *aurora borealis*, or the Northern Lights.

Below: The Earth's magnetism produces an invisible force field that can be felt everywhere. A magnet, or a compass needle, will always be pulled to point toward the magnetic North Pole.

The Earth is just bursting with energy. We can see some of it on the surface, as waves and tides crash away at the seashore, and as hurricanes and gales sweep the land. But deep down within the Earth, huge pressures and very high temperatures create just as much violent activity.

One way in which we can see this is in the Earth's magnetic field.

THE BIG MAGNET

If we look at an ordinary magnetic compass, we will see that the needle always swings around so that one end points to the north. That is because the whole Earth is acting like a great big magnet. The compass needle itself is a magnet and, as with all magnets, one end is attracted to a magnetic pole and the other end is repelled.

The Earth's magnetic field is probably caused by the fact that the solid core of the Earth is spinning at a different rate from the rest of the planet. It can do this because the outer core is liquid. The energy of this spinning produces the magnetic field, in the same way as magnetism is involved in the turning of an electric motor.

THE SHAPE OF THE FIELD

The two points at which the Earth's magnetism is strongest are called the poles. These can be found on the Earth's surface by looking at which way compass needles point. The magnetic poles are always quite close to the North and South geographic Poles, but they move about all the time. For this reason, maps that show the direction of the magnetic poles have to be updated every few years.

The effect of the magnetic field reaches far out into space. Space scientists must take this into account when they prepare sensitive instruments for use in satellites.

Left: Animals can sense the magnetic field and use it to navigate over long distances. Migrating salmon use it to find their way back to their own rivers.

MOVEMENT OF THE CRUST

The Earth has two types of crust. There is the continental crust, that we see all the time as the land, and there is the oceanic crust that is usually hidden away deep on the bottom of the sea. These crusts were formed by quite different processes.

The surface of the Earth is moving about, and constantly being renewed, all the time. This process is known as "plate tectonics." Imagine that the surface of the Earth is made up of a number of plates, like the panels of a soccer ball. Imagine also that each plate is growing along one seam. At the same time, the plate is moving away from that seam and being destroyed against the seam at the other side. That is more or less what is happening on the Earth's surface all the time.

MOVING PLATES
Each plate consists of the crust – oceanic and continental – and the topmost layer of the mantle. It rides on the soft layer of the mantle beneath.

The seams along which the plates grow are rarely seen. They lie at the bottom of the oceans. Here, hot material from the mantle wells up as a series of underwater volcanoes forming a ridge along the ocean floor. The molten material becomes solid and produces new crust and mantle. Then this cracks and moves apart as even newer material wells up in between. We know this happens because the rocks by the oceanic ridges are quite new, and they become older as they move farther away from the ridge crests.

The plates are being destroyed at the other side, where they are pressing up against their neighbor. When this happens, one plate is ground down beneath the other and melts as it returns to the mantle. This happens along the deep ocean trenches. The volcanic island chains of the Pacific Ocean show where plates are being destroyed.

Above: Slowly, slowly the surface of the planet is moving. New crust churns up along ocean ridges and spreads outward, with the continents carried around like rafts.

Left: In Iceland it is possible to see the plate edges. A vast crack separates the North American plate on the left, from the Eurasian plate on the right.

Below: The Earth's surface consists of about half a dozen major plates and some smaller ones. Each plate is being created at one edge and then moves away to be destroyed at the other edge.

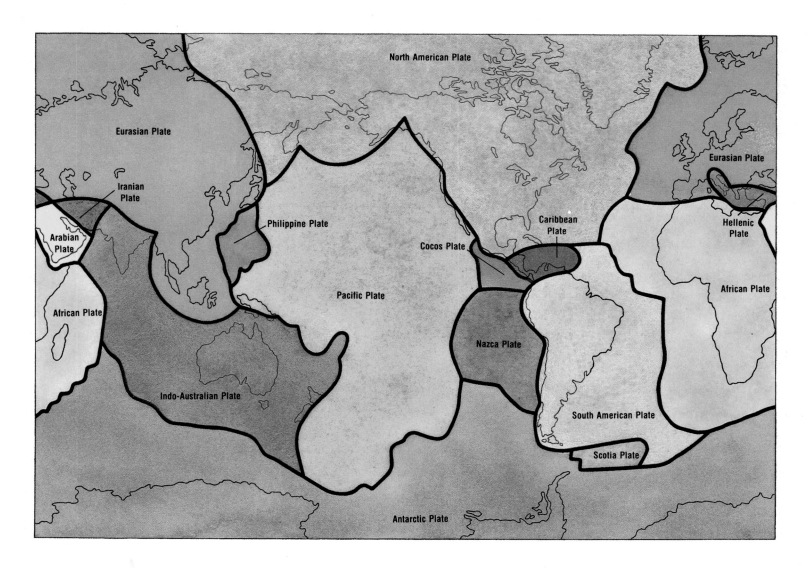

North American Plate

Eurasian Plate

Iranian Plate

Philippine Plate

Arabian Plate

Caribbean Plate

Cocos Plate

Eurasian Plate

Hellenic Plate

African Plate

African Plate

Pacific Plate

Nazca Plate

Indo-Australian Plate

South American Plate

Scotia Plate

Antarctic Plate

CONTINENTS

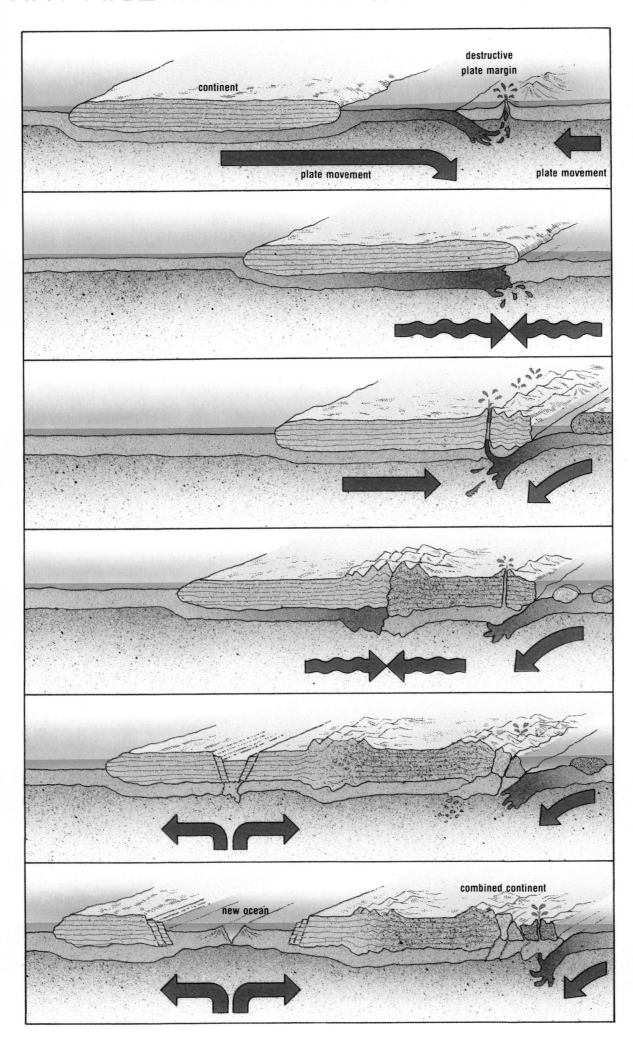

Left: Every continent has a dramatic history. Let us start with a mass of continental material being carried toward a destructive plate margin.

When it hits the plate margin, where the plate is being pulled down and destroyed, the system jams. The continent is too light to be pulled down, too.

Pressures build up, and finally the other plate begins to be pulled down beneath the continent. The continent crushes up into mountains at the edge, shot through by volcanoes formed from melting plate material.

Another continent brought along by the second plate collides with the first. The movement that is pulling it downward stops again and restarts at the edge of the new, combined continent.

New continental pieces continue to be added. Later, new tensions may develop, and the continent begins to pull apart, forming a rift valley over rising mantle material.

The continent splits in two as the movement continues. An ocean grows between the pieces. Thus, the geography of the world is constantly changing.

The oceanic crust is constantly being created and destroyed – nowhere is it more than about 200 million years old. The continental crust, however, has been around for a much longer time.

The continents are made of a lighter material than the ocean floors, and are carried about embedded in the surface plates like logs embedded in ice. As they are light, they cannot be drawn down into the mantle where the plates are destroyed. The continents remain on the surface and crumple up into mountain ranges while the ocean crust is being swallowed up nearby.

RECYCLED CRUST

The continental crust is made of remelted crust and is continually growing. When an oceanic crust is drawn down into the mantle and melted, some of the molten material rises again to the surface and forms volcanoes and island chains. Sometimes these chains become joined to the edges of the continents, making them bigger.

When two plates, each carrying continents, collide with one another, the two continents fuse to form a big supercontinent. The line of the join is marked by mountain ranges. The modern Himalayas show where the old continent of India collided with Asia about 50 million years ago.

JIGSAW CONTINENTS

Sometimes a new plate boundary appears beneath a continent, and the plates begin to move away from it. When this happens, the continent tears itself apart and forms two new land masses. This is happening along the Great Rift Valley of East Africa today. It is also why the east coast of South America matches the west coast of Africa. These two once formed a single continent, but about 100 million years ago they split along the middle, producing the Atlantic Ocean in between.

Below: Along the Great Rift Valley of East Africa, the continent is beginning to split apart. The whole eastern part of Africa may one day form a new continent in the Indian Ocean.

BECKY C. LOUCKS   october 10. Sunday 94

VOLCANOES

The runny lava from a basaltic volcano builds up a broad, low shield volcano.

Left: Basaltic lava squirts up as fire fountains and then flows as glowing molten rivers. Basaltic lavas are tourist attractions as shown here, in Hawaii.

The surface plates of the Earth are constantly being created and destroyed. Violent energy is produced while this is going on. When this energy is released at the surface of the Earth, volcanoes are produced.

As the plates are being created on one side and destroyed on the other side, there are two different types of volcano, one at each place. Both types of volcano are produced by molten material from within the Earth, which is called magma. This changes as it rises through the crust, and the liquid that remains erupts at the surface in the form of lava.

GENTLE ERUPTIONS

Where plates are forming, the lava comes from deep in the mantle. This kind of lava is called basalt, and is black and runny. At the surface it pours and flows for a long way before becoming solid. As a result the volcanoes are broad and flat. Often the lava flows from a crack rather than a hole, or the summit of a volcano may collapse to form a vast pool of molten rock. An eruption of this kind produces glowing fountains.

KILLER VOLCANOES

On the other hand, the volcanoes that erupt where the plates are being destroyed are very violent and dangerous. Here the magma consists of molten crust. The lava that is produced, called andesite, is much stiffer and stickier. Tall, steep-sided volcanoes form and very often the lava solidifies in the vent, stopping up the eruption. When this happens, the pressure beneath builds up so much that the whole volcano eventually explodes, devastating the countryside for miles around.

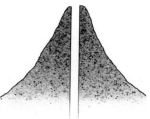

The thick lava from an andesitic volcano does not flow far. It builds up a steep-sided volcano.

Left: Sticky lava may jam the funnel of an andesitic volcano, causing disastrous explosions. Such volcanoes, like Mount Saint Helens, are treated with respect.

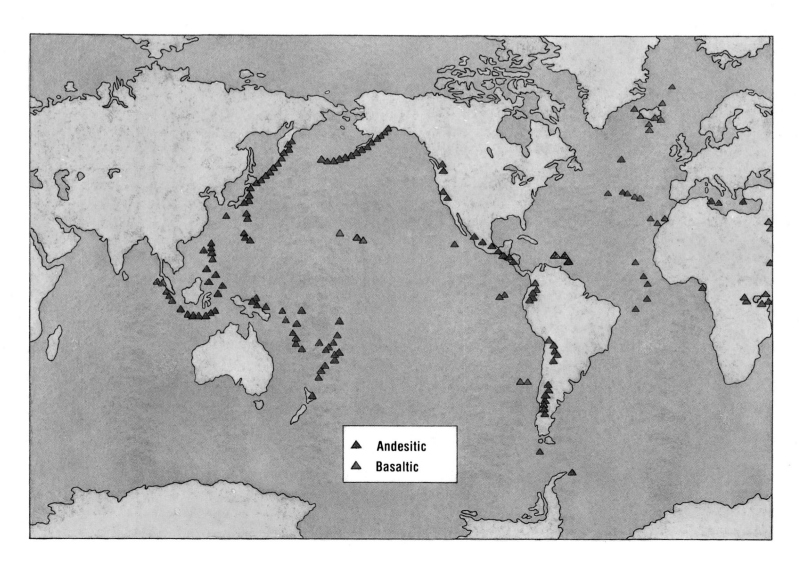

▲ **Andesitic**

▲ **Basaltic**

EARTHQUAKES

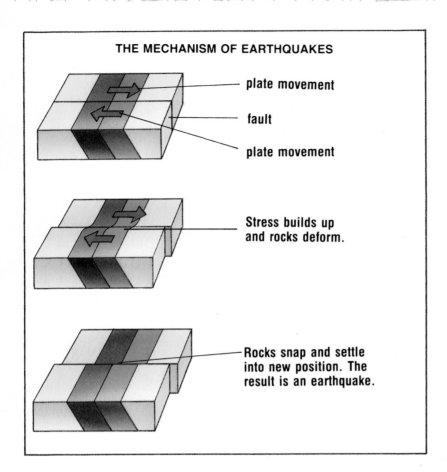

THE MECHANISM OF EARTHQUAKES

plate movement

fault

plate movement

Stress builds up
and rocks deform.

Rocks snap and settle
into new position. The
result is an earthquake.

A cross the surface of the Earth, the moving plates rub against one another. They set up strains and tensions as they pull and push, and eventually the rocks give way and crack. An earthquake results.

EARTHQUAKE ZONES
It is hardly surprising that most earthquakes are felt along the edges of the plates. People who live in such places as Japan, or California, or the Mediterranean area, can expect earthquakes frequently. However, deep-seated earthquakes may occur in places a long way from plate margins, such as the British Isles. This shows how little we really know about the movements deep in the crust.

In an earthquake, the rocks move along a crack, called a fault. The point where the greatest movement occurs is called the focus, and the point on the surface immediately above is called the epicenter. Most damage is usually caused at the epicenter. Vibrations spread out from the focus and eventually die out, absorbed by the rocks.

The sudden movement usually carries the rocks too far. Within hours or days, they may jump back to settle and cause an aftershock.

MEASURING EARTHQUAKES
We can measure an earthquake in two ways. First we can estimate the total amount of energy released by

SHOCK WAVES

epicenter

fault

focus

Above: The focus is where most of the earthquake movement takes place. Shock waves spread out from it. The epicenter is the point on the surface directly above the focus.

Left: Horrific news pictures such as this of Mexico City in 1985 confirm the destructive power released by an earthquake.

THE MERCALLI SCALE

1. Only felt by instruments.
2. You feel it if you are upstairs. Light bulbs swing.
3. You feel the earthquake indoors.
4. You feel it outdoors. If you are inside it feels like a truck hitting the building, making windows rattle.
5. It wakes you up if you are asleep, and cracks the plaster.
6. Windows and plates break, and plaster falls.
7. It is difficult to stand and loose bricks fall.
8. Chimneys fall down and you can't steer a car.
9. The ground cracks and buildings collapse.
10. All houses are smashed and landslides start.
11. Railway lines bend and bridges and underground pipes are destroyed.
12. You would be lucky to survive this one. Objects are thrown into the air and everything is destroyed.

the earthquake itself. This is called its magnitude and is measured on the Richter scale. Each point on the scale represents ten times the amount of energy at the point lower down. The strongest earthquakes measure about point 8 on the Richter scale.

Another way to measure earthquakes is by the intensity, and this is based on the effects at different places. The intensity is measured on the Modified Mercalli scale. At point 1 on the scale, the effect can only be detected by sensitive instruments, while at point 12 there is violent movement, and buildings are destroyed.

Any earthquake will have just one measure on the Richter scale, but its estimates on the Modified Mercalli scale will vary from the epicenter to places farther away where less damage will occur.

MINERALS

Look at a rock, any rock, through a magnifying glass or a microscope. You will see that it is made up of a mass of small fragments. These are pieces of chemical substances that we call minerals.

Every rock is made up of, perhaps, half a dozen different kinds of minerals. There is a different set of minerals in each type of rock. Hundreds of different minerals exist, but there are only a few that occur time after time in the rocks of the Earth's crust.

Right: When a thin slice of rock, like this granite, is viewed under polarized light through a microscope, the different minerals show up in fantastic colors.

Right: A gold nugget is a "native ore" – one that contains the metal and nothing else, unlike iron ores which are mostly iron oxides. Garnets sometimes form good crystals in metamorphic rocks.

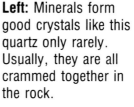

Left: Minerals form good crystals like this quartz only rarely. Usually, they are all crammed together in the rock.

THE ROCK-FORMING MINERALS

The most common chemical substance in the earth is silica, which is like glass. Many metals can combine with this silica to form what we call the silicate minerals. The semi-precious stone garnet, which looks like colored glass, is a silicate mineral containing aluminum. Olivine is another silicate mineral, this one containing iron and magnesium. Rocks are made up mostly of silicate minerals like these.

Since these common minerals contain iron, magnesium and other metals, we might think that they would be valuable storehouses of these metals. Unfortunately, when the metal is combined with silica it is almost impossible to remove.

THE ORE MINERALS

The minerals from which we can extract the metal are much rarer. They are called ore minerals and may consist only of the metal itself, such as gold nuggets. Sometimes the metal may be combined with oxygen, such as the various iron ores, or with sulfur and oxygen, in minerals called sulfates. These are also good ore minerals. Only when the geological conditions have concentrated large masses of these minerals together, however, is it worthwhile mining them.

IGNEOUS ROCKS

Right: Igneous rocks form from molten material forced into the rocks of the crust or blasted out at the surface. They form distinct structures underground.

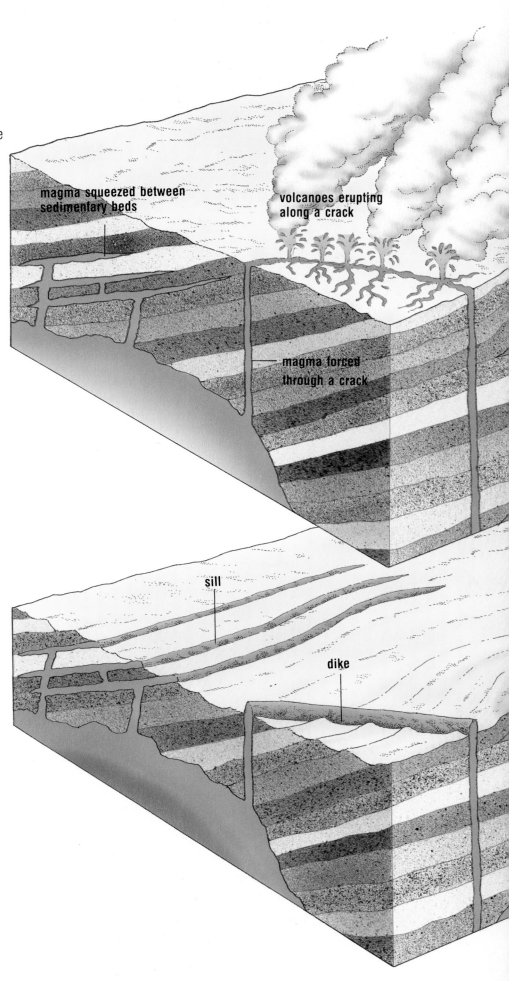

magma squeezed between sedimentary beds

volcanoes erupting along a crack

magma forced through a crack

sill

dike

R ock is the name that we give to the substance of the Earth's crust. We know of three ways in which rock can form.

HOT ROCKS
Perhaps the most straightforward type of rock is the igneous rock. This forms when molten material from the Earth's interior cools and becomes solid.

We can actually see certain kinds of igneous rock forming. As lava erupts from a volcano, it pours down the mountainside, cools and then hardens. This, however, is only one type of igneous rock. The kind of rock formed in this way is called an extrusive rock, because it is squeezed out at the surface of the Earth. Basalt and andesite are typical extrusive rocks.

The other kind of igneous rock is called intrusive rock. This forms when the molten material becomes solid underground without reaching the surface. We do not usually see these igneous rocks until millions of years after they formed, when the rocks that lie above them have been

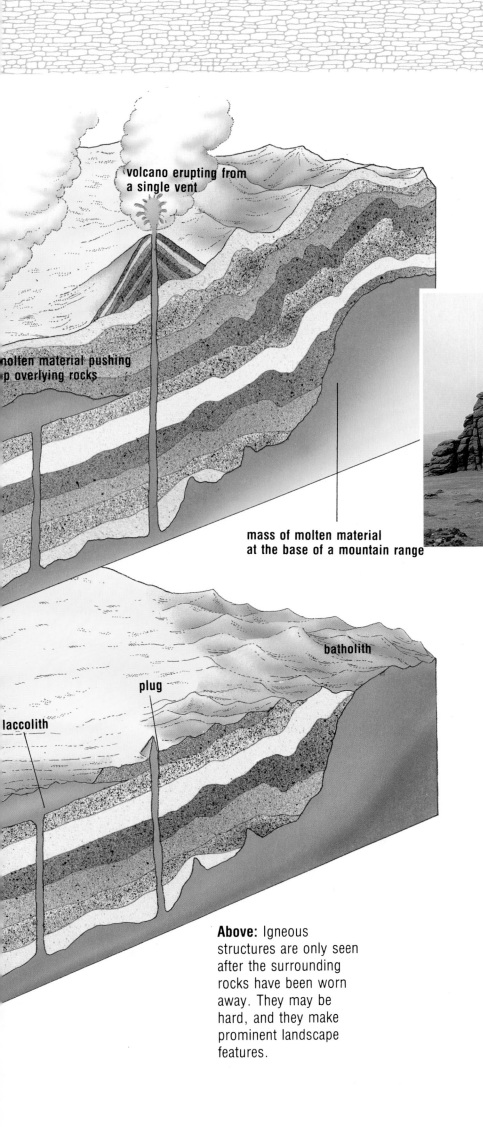

volcano erupting from
a single vent

molten material pushing
up overlying rocks

mass of molten material
at the base of a mountain range

batholith

plug

laccolith

Above: Igneous
structures are only seen
after the surrounding
rocks have been worn
away. They may be
hard, and they make
prominent landscape
features.

Below: Exposed granite
erodes along cracks and
joints, breaking up the
mass into a castle-like
structure called a tor,
such as this on
Dartmoor, in England.

worn away. An intrusive igneous
rock takes longer to cool than an
extrusive one – all the rocks around
it insulate it and keep it hot – and
so the minerals in it have time to
form large crystals.

MANY MAGMAS, MANY ROCKS
Different mixtures of molten
material – or magma – will produce
different types of igneous rock.

For example, a magma that does
not have much silica will produce
rock called gabbro if it cools slowly
in large masses. It will produce rock
called dolerite if it cools more
quickly in cracks in the crust. If it
erupts at the surface as lava, it will
produce rock called basalt. On the
other hand, a magma rich in silica
will produce coarse granite in large
masses formed underground, and
extrusive andesite if it erupts as lava
at the surface.

SEDIMENTARY ROCKS

Huge areas of the Earth's surface are covered with sedimentary rock. This forms when millions of particles build up in layers, usually on the bottom of the sea, and these layers are later compressed and cemented together. Scientists distinguish between three main types of sedimentary rock.

ROCKS FROM RUBBLE

One type forms when rocks that already exist are broken down by the wind and the rain. The broken fragments are washed down to the sea by rivers and streams and are deposited as sand or mud. This kind of rock is called a clastic sedimentary rock.

The finest fragments form mud, which eventually produces mudstone and shale. Larger fragments produce sand, which eventually forms sandstone. Even larger pieces form rubble that may eventually produce the coarse rocks called conglomerate and breccia.

ROCKS FROM LIFE

Another type of sedimentary rock is called biogenic sedimentary rock. This is made up of pieces of material that were once living things. Corals build up reefs of solid shell material, and the shells of millions of microscopic swimming creatures can sink to the bottom of the sea. Both of these eventually become the biogenic sedimentary rock called limestone.

ROCKS FROM CHEMICALS

The last type of sedimentary rock is the chemical sedimentary rock. Seawater is salty. If a lagoon of seawater dries up in the hot sun, the salt is left behind as a layer of rock salt – a chemical sedimentary rock. Sometimes currents bring lime-rich seawater up from the deep oceans, and the lime crystallizes out in the shallows to form another of the kinds of limestone.

We can recognize a sedimentary rock when we see it because it usually forms distinct layers that are called beds.

rocks eroded by wind

salt deposits in desert lakes

desert sands blown by wind

shingle on beaches

collection of limy fragments in coral reefs

Above: Sedimentary rocks are formed from sediments. These deposits accumulate in different ways, but are mostly derived from debris broken and eroded from earlier rocks.

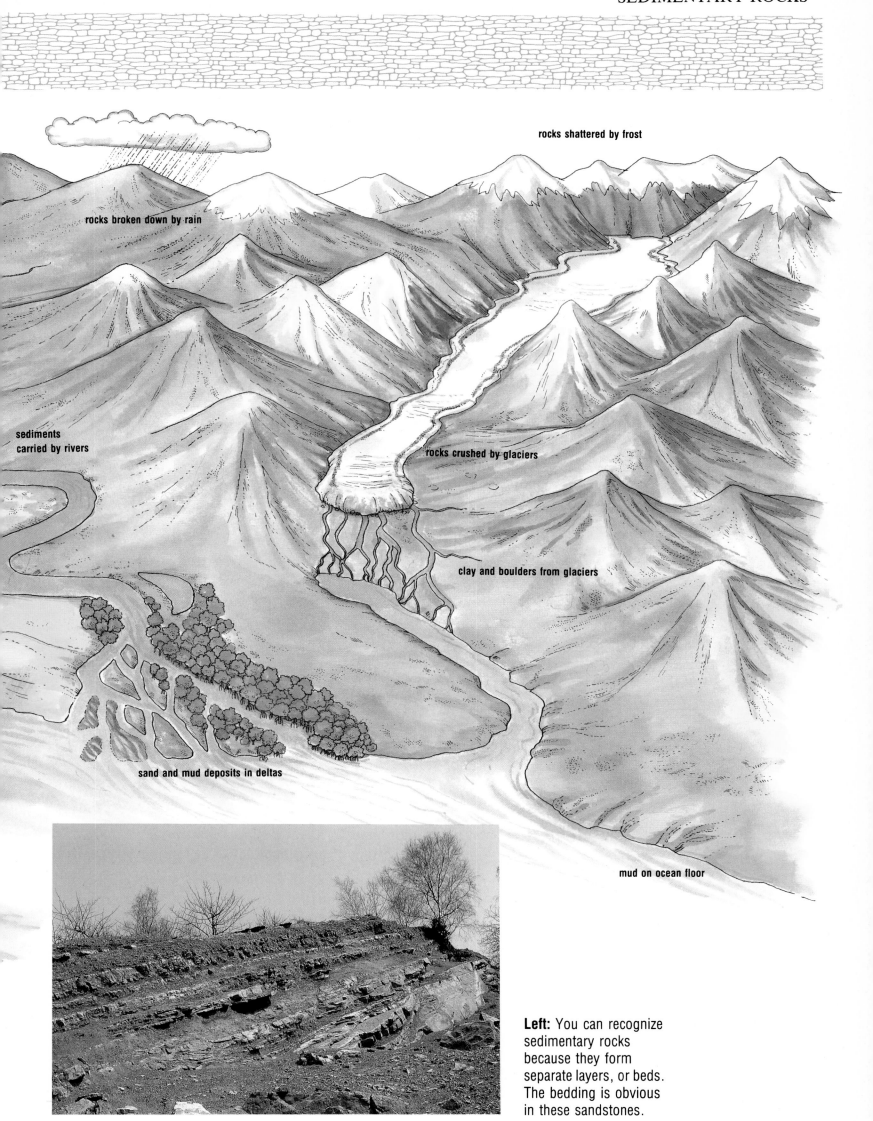

rocks shattered by frost

rocks broken down by rain

sediments
carried by rivers

rocks crushed by glaciers

clay and boulders from glaciers

sand and mud deposits in deltas

mud on ocean floor

Left: You can recognize
sedimentary rocks
because they form
separate layers, or beds.
The bedding is obvious
in these sandstones.

29

METAMORPHIC ROCKS

Right: Rocks that have been altered by pressure often have their minerals aligned, so that the rock splits easily, as in these slates.

Below: Pressure on a rock alters its mineral content, and different metamorphic rocks are produced as the pressure increases in the depths of a mountain chain.

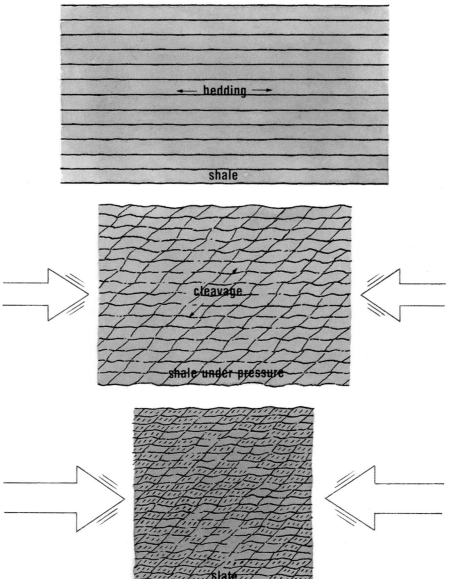

The third way in which a rock can form is by conversion from some other kind of rock. The rocks formed this way are called metamorphic rocks, meaning rocks that have changed.

When a continent lies at the edge of one of the plates of the Earth's crust, and is being crushed and crumpled by another plate grinding down against it, the forces produced are immense. It is these forces that produce the change, or metamorphism.

FORMED BY PRESSURE

Deep in mountain chains, the minerals and rocks are subjected to the most incredible pressures. These may be so great that one kind of mineral changes into another. The new rock produced by this pressure is called a regional metamorphic rock. Sometimes the new minerals are lined up in one direction, enabling the rock to split easily into thin sheets, such as can be seen in the regional metamorphic

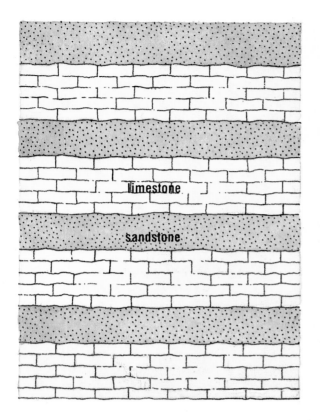

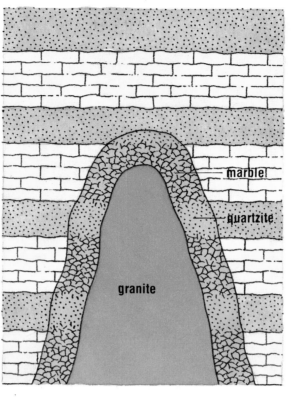

Left: Heat from a cooling igneous structure cooks the surrounding country rocks, making new minerals and forming a metamorphic rock.

Below: Rocks that have been altered by heat tend to be massive and have no internal structure, like this marble. It is used in sculpture because it will not split.

rock called slate.

More intense pressure will crumple the sheets of flat mineral and produce a wavy effect, as in the rock called schist. Under the greatest pressures, different minerals will form in different bands, producing the rock called gneiss.

FORMED BY HEAT

Another type of metamorphic rock is called thermal metamorphic rock. In this type, the minerals are changed by heat rather than by pressure. These rocks form in much smaller masses than the regional rocks. They are usually found only alongside intrusive igneous rocks, where the rocks around it have been cooked by the heat of the magma mass.

The important thing about a metamorphic rock is that all the changes were produced while the rock was still solid. If the rock melted and solidified again, the result would be an igneous rock.

FOLDS AND FAULTS

The movement of the Earth's crust leaves its scars. Rocks are twisted and broken as the continents grind against one another. Sometimes this damage is caused quickly, resulting in earthquakes, but more often the rocks are distorted in a slow, creeping movement that would not be noticed.

FOLDS
When the rocks bend, the result is known as a fold. Folds can be seen in mountainous areas, where engineers have cut routes for roads and rail tracks through the rock. When the mountains were thrown up, the rocks were squeezed from each side. Flat-lying beds crumpled up into a wavy pattern like a tablecloth does when it is pushed along a table.

A fold that arches upward is called an anticline, while one that sags downward is called a syncline. These are usually found together, following one another.

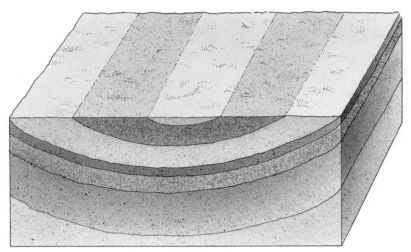

syncline

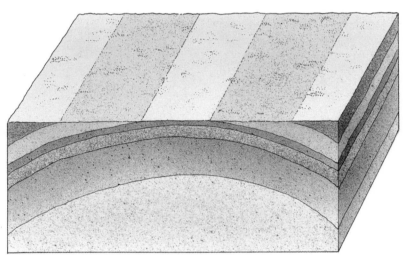

anticline

Above: A sagging fold in the rock, with the rocks dipping inward, is called a syncline. One that arches upward is an anticline.

Left: Folds, such as these on a beach in southwestern England can only be seen once the overlying rocks have been worn away.

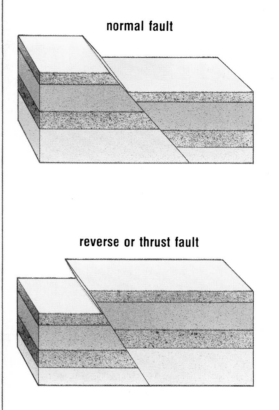

normal fault

reverse or thrust fault

horst

graben

FAULTS

When the rocks crack, and the masses at each side of the crack move in relation to one another, the result is a fault. Most earthquakes occur as rock masses move along a fault. Again, different names are given to different types. If the rock appears to have moved downward along the fault plane, then the fault is a normal fault. If it has moved upward, it is a reverse fault.

A normal fault is caused by the rocks being pulled apart. Often a block moves downward between two or more faults, giving a structure called a graben. A block that is left standing while the rocks at each side have moved downward is called a horst. Reverse faults are caused by compression, like the forces that cause folds to shear.

Above: A fault is usually seen as a crack in the rocks along which the rocks have shifted. It is often possible to see which way the rocks have moved.

Right: Normal faults, reverse faults, horsts, and grabens are formed by forces acting on the rocks. These forces compress them from the side or pull them apart.

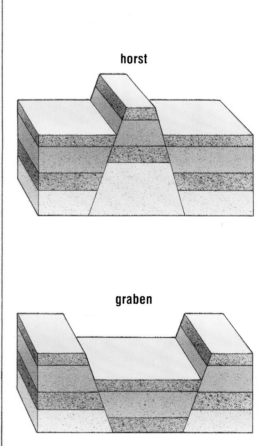

33

MOUNTAINS

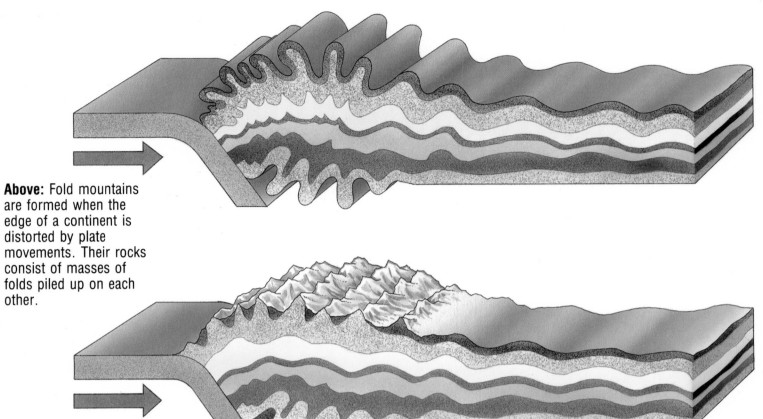

Above: Fold mountains are formed when the edge of a continent is distorted by plate movements. Their rocks consist of masses of folds piled up on each other.

Above: In reality, the tops of fold mountains are eroded as fast as they are thrown up. The mountain crests do not take on the rounded appearance of the folds.

Right: The Himalayas – the greatest mountain range on Earth – consist of rocks that have been folded over the past 50 million years.

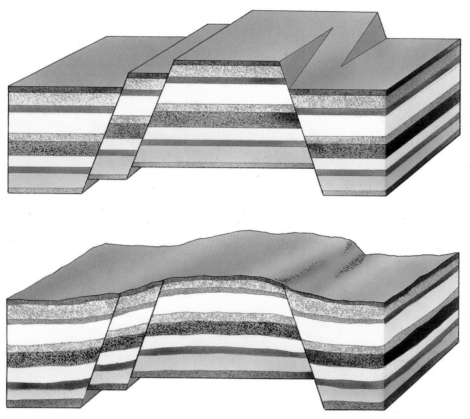

Left: Block mountains, formed by faulting, do not often show the sharp edges suggested by fault movements. They are usually worn away into rounded shapes by erosion.

One of the most spectacular sights of the scenery on our planet is the mountains. These are produced, like the folds and faults, by the movements of the plates and of the continents. They are, in fact, the results of folding and faulting on a vast scale.

FORMED BY FOLDING
The greatest ranges of mountains on Earth are fold mountains. Those like the Andes of South America are being formed today as the edge of the continent is being crumpled by the pressure of an oceanic plate that is plowing down beneath it. The Rockies of North America were formed in the same way, although they have stopped growing because the activity now lies in the Coast Ranges of the new continental edge.

Even greater are the Himalayas. These began as a mountain range like the Andes along the southern edge of the Asian continent. Then along came the old continent of India and crashed into it, thrusting up the greatest mountain range on Earth. The Ural Mountains between Europe and Asia formed in this way some 300 million years ago. Now they are worn down by wind and rain to mere stumps of their former selves.

FORMED BY FAULTING
Faults produce mountain ranges as well, but these are not so extensive as the fold mountains. They tend to form where continents are beginning to tear apart as new plate boundaries appear beneath them. The great pressure from below thrusts upward, cracking the crust into blocks and moving them up and down. Along the Great Rift Valley of Africa, there are mountains which were formed in this way. The highlands at the edge of the Red Sea were also formed like this.

EARTH'S TIME SCALE

The Earth is thousands of millions of years old – 4½ billion, in fact. But can you imagine such a time span? Can you imagine even a million? Geologists found that hundreds of millions of years, and thousands of millions of years, were very clumsy numbers to work with, and so they split up the history of the Earth into different eras and periods, depending on what was going on in the world at that time.

DARK AGES

The first seven-eighths of geological time – the part that we do not know a great deal about because it is so old – is called the Precambrian. This covers the time from the formation of the Earth to about 590 million years ago. We know of very few fossils from this time. The first part of the Precambrian is called the Archaean – when there was no life on Earth. The rest is the Proterozoic – when life of some sort existed.

AGES OF ENLIGHTENMENT

Then, about 590 million years ago, at the beginning of the Cambrian period, we begin to find many fossils. Animals suddenly developed hard shells which were easily preserved. From then on the history of the world is easy to read from the rocks. This whole segment of time is called the Phanerozoic – the time of obvious life.

The Phanerozoic is divided into three eras. These are the Paleozoic, the time of ancient life; the Mesozoic, the time of middle life; and the Cenozoic, the time of modern life.

These eras are then subdivided into periods according to what animals and plants lived at the time.

Era	Time Periods	Millions of Years Ago
Cenozoic	Quaternary	2
Cenozoic	Tertiary	65
Mesozoic	Cretaceous	144
Mesozoic	Jurassic	213
Mesozoic	Triassic	248
Palaeozoic	Permian	286
Palaeozoic	Pennsylvanian	320
Palaeozoic	Mississippian	360
Palaeozoic	Devonian	408
Palaeozoic	Silurian	438
Palaeozoic	Ordovician	505
Palaeozoic	Cambrian	590
	Precambrian	

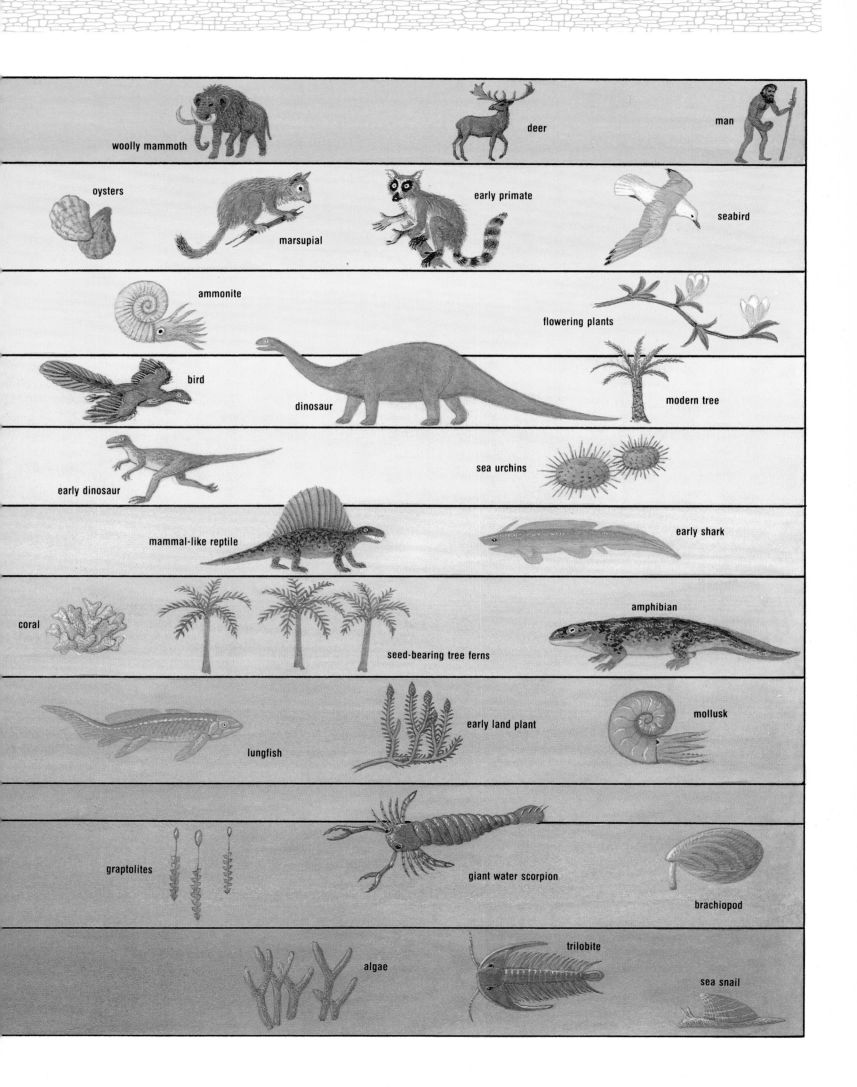

woolly mammoth

deer

man

oysters

marsupial

early primate

seabird

ammonite

flowering plants

bird

dinosaur

modern tree

early dinosaur

sea urchins

mammal-like reptile

early shark

coral

seed-bearing tree ferns

amphibian

lungfish

early land plant

mollusk

graptolites

giant water scorpion

brachiopod

algae

trilobite

sea snail

READING EARTH'S STORY

A geologist can look at the layers, or beds, of sedimentary rock and work out from them what conditions were like in ancient times.

The nature of the rocks themselves is important. If we see limestone, we can work out that there was once a clear sea in the area. Shale means that the rocks were formed under muddy waters. Sandstone tells us that there were great deposits of sand in the region, on a seashore, at the bottom of a river, or in a desert.

TELLTALE STRUCTURES
In the example of the sandstone, a close look at the rock will tell us where it was formed. Sand dunes leave a characteristic curved bedding in the rocks. River currents produce the same kind of S-shaped bedding, but on a much smaller scale. Sand formed on beaches often has small ripple marks caused by the waves.

Ocean currents also leave their marks in the sediments, scooping out hollows and leaving the debris elsewhere. A river emptying into the sea leaves heavy sand first and the lighter material later. A rock formed this way shows a graded bedding, with coarse material at the bottom and finer at the top.

Below: Once upon a time there was a desert here. We can tell that because of the shapes of the sand dunes that are left in the rocks.

Where water dries out, the mud beneath cracks into polygonal shapes that can be preserved in the rock.

THE GEOGRAPHY OF THE PAST

The study of these rock types and rock structures is the science of stratigraphy. The clues that are found can help geologists to build up a picture of what the land was like in times past. They can find out where the rivers and seas were, and even what the climate was like and what animals and plants lived there. This study is known as paleogeography (*paleo* is the Greek word for ancient).

Above: A river current builds the sand out into tongues of sediment. Millions of years later these tongues are visible as curved beds in the rock.

Right: Waves washing back and forth on a seashore throw the sand into ripples. These can be preserved in rock and will show where there was once shallow water.

FOSSILS AND EVOLUTION

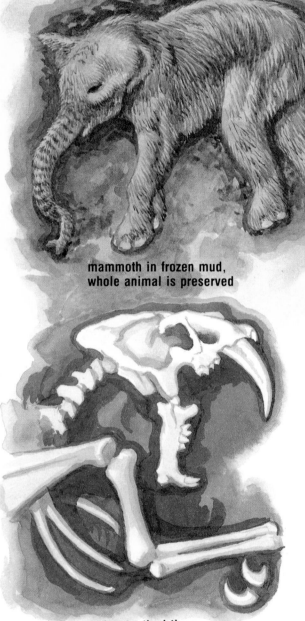

mammoth in frozen mud,
whole animal is preserved

saber-toothed tiger,
skeleton bones are preserved

Fossils form in
many different ways.
Usually, the living matter
has been replaced by
minerals.

Below: A fossilized fern
leaf is made from the
black carbon in the
original plant material.

Perhaps the most spectacular evidence of ancient conditions can be seen as the remains of animals and plants that lived in former times. We call such remains fossils.

The different beds of a rock sequence have different fossils in them. This is partly because the different rocks were formed in different environments – for example, you would not find the same animals living on a muddy sea bottom as you would on a sandy riverbed. Another reason is that animals and plants have changed throughout time.

THE MARCH OF LIFE

The very earliest animals – back in Proterozoic times – had small, soft bodies and left few fossils. Then, at the beginning of the Paleozoic, hard shells developed. Shortly afterward the first vertebrates – the backboned animals – appeared. They were primitive fish. These were all sea creatures, because until this time the atmosphere on Earth was poisonous.

Plants first colonized the land, followed by insects, and then, halfway through the Paleozoic, the fish. Over millions of years a group of land-living fish became the first amphibians – like today's frogs and salamanders – and from these the reptiles developed. The reptiles were fully land-living vertebrates.

petrified forest,
wood is replaced by minerals

dinosaur footprint,
only the track is preserved

HOW A CAST IS MADE

A shellfish lives on the seabed.

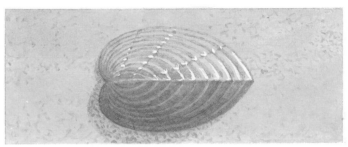

The shell of the dead animal is covered by sediment.

The sediment turns to rock, and the shell decays to
leave a cavity – a mold.

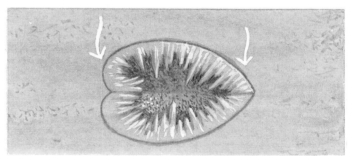

The mold fills with minerals deposited by ground water.

The mineral lump, in the shape of the original shell,
is called a cast.

In Mesozoic times the reptiles became really important, but some of them changed into mammals and birds as well. However, the mammals and birds were not nearly as important as the great dinosaurs of the time.

Then, at the end of the Mesozoic, the big reptiles died out and the mammals and birds took over. Throughout the Cenozoic these were changing all the time too, and eventually, just about 2 million years ago, one group of mammals developed into humans.

41

THE WATER CYCLE

Our planet is unique in the Solar System because of the presence of water. The normal temperatures and pressures found at the Earth's surface mean that water can exist in its three forms – as solid, as liquid, and as gas or vapor. The liquid water makes life possible, and the movement of water gives us environments that are suitable for sustaining life.

Most water exists as liquid in the oceans. From the surface it evaporates into the atmosphere as vapor or clouds. When the vapor cools, it condenses into liquid again and falls as liquid rain, or as solid hail or snow. If it falls on the land, it runs off as rivers and returns to the ocean. This continuous process is called the water cycle.

The water cycle is not quite as neat and simple as is suggested here. There are all sorts of side loops as well. Much of the water that falls on the land soaks into the soil and rocks and becomes ground water.

THE WATER TABLE

There is a certain level beneath the surface where the rocks and soils are thoroughly soaked and contain as much water as they can hold, like a sponge that is full of water. This is called the saturation zone. The upper limit of this zone is known as the water table. The water table may be close to the surface in damp climates, or deep down in desert areas. Where the water table actually reaches the surface, as in a hollow or a side of a hill, the water seeps out and forms a spring.

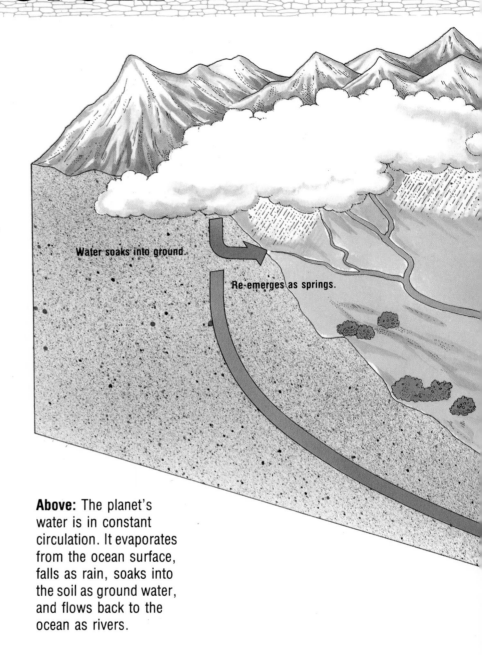

Water soaks into ground.

Re-emerges as springs.

Above: The planet's water is in constant circulation. It evaporates from the ocean surface, falls as rain, soaks into the soil as ground water, and flows back to the ocean as rivers.

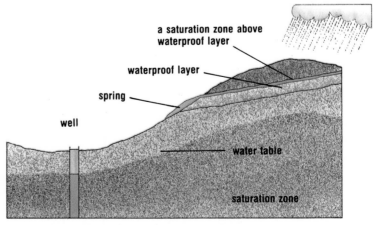

THE WATER TABLE

a saturation zone above waterproof layer

waterproof layer

spring

well

water table

saturation zone

Springs form where the water table meets the surface naturally. Wells for drinking water are dug down to reach into the saturated zone beneath the water table.

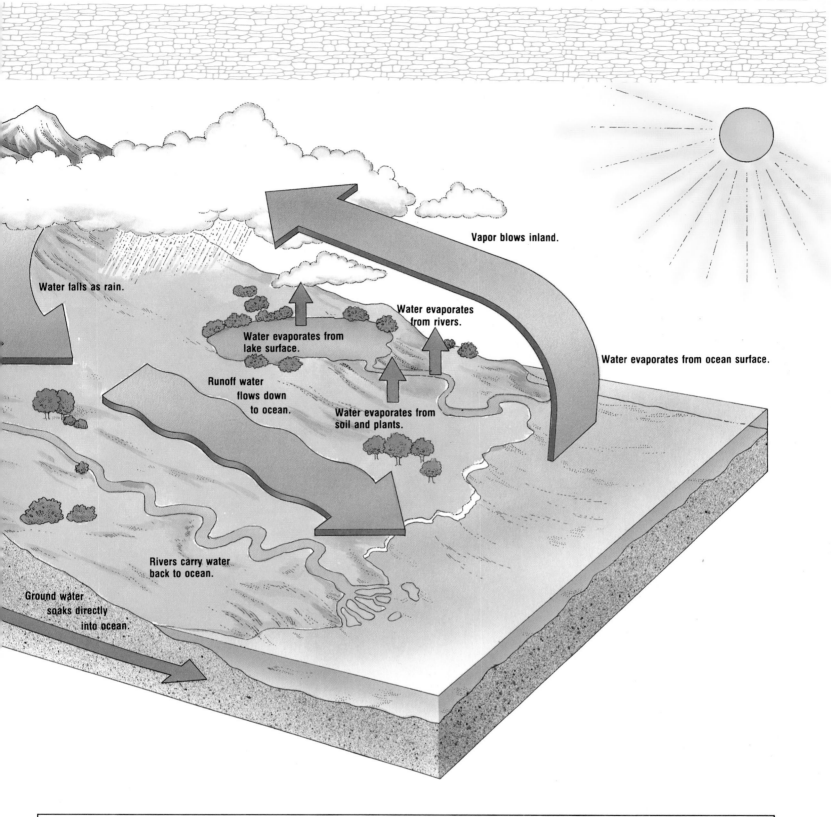

Vapor blows inland.

Water falls as rain.

Water evaporates from lake surface.

Water evaporates from rivers.

Water evaporates from ocean surface.

Runoff water flows down to ocean.

Water evaporates from soil and plants.

Rivers carry water back to ocean.

Ground water soaks directly into ocean.

ROTTING ROCKS

Rain is not pure as we might think. As it falls, it dissolves the gas, carbon dioxide, from the atmosphere and becomes a weak acid. When this water seeps into the ground, it can react with certain minerals there. The mineral calcite, that forms limestone, is particularly vulnerable. Some limestone areas have huge caves dissolved away by the slow movement of ground water. The calcite that is dissolved to make the caves is redeposited as stalagmites and stalactites.

RIVERS

Above: A youthful river splashes vigorously down its valley, forming waterfalls and rapids. It easily wears away its bed and cuts deep gorges in mountainous areas, as here in Norway.

Rain that falls on the ground flows over the surface and into streams, or it sinks into the soil to become ground water and rises again in springs. Either way it ends up flowing downhill toward the sea. Small streams unite into larger streams and eventually these become rivers.

The rivers we see can be swift, splashing and full of energy; or they can be steady and purposeful; or they can be slow and weak. These represent the three stages of river formation as recognized by geographers.

THE THREE STAGES
In its youthful stage the river is full of strength. Its erosive power cuts downward into its bed, forming deep valleys and gorges. Stones and

44

debris are ripped up and carried along. Waterfalls and rapids are common in young rivers.

Then it passes into its mature stage. Here the water moves much more slowly, but it is still quite strong. It wears away its bed and its banks, but it also deposits some of the debris it carries along. Mature rivers meander in broad valleys.

Finally the river passes into old age. It moves slowly in huge loops, depositing material as it goes. It is now too weak to do any eroding at all. There is no valley here, just a plain built up of sand and mud deposited by the water before it reaches the sea.

Above: A mature river flows lazily in a broad flat valley, as here in France. It wears back the sides of the valley, but deposits silt and sand on the floor.

Left: An old-age river creeps along sluggishly, changing its course all the time and dropping all the sediment it brings from the far mountains.

GLACIERS

When water cools to 32°F, it passes from its liquid phase into its solid phase – it becomes ice. We see it in our refrigerators and on puddles on cold mornings, but in many places of the world, the ice is an important and permanent part of the natural landscape.

MOUNTAIN ICE

In high mountains the snow that falls all the time builds up in valleys without melting. When layer piles upon layer, the snow at the bottom becomes squeezed into ice – just like when you squeeze snow to make a snowball and it forms a hard icy lump. The mass of valley ice formed this way moves slowly downhill and forms a glacier.

When ice is compressed at the base of a glacier, it can be made to move like putty does when you mold it in your hands. That is why a glacier can travel down the valley floor. The ice on the surface is not under pressure and so it is brittle. It cracks and snaps into crevasses and jagged pinnacles as the glacier moves onward. The immense weight tears up the valley's floor and sides, widening it into a broad U-shape. When the glacier eventually melts, all the debris is left behind in mounds of clay and rubble that are called moraine.

ICE AT THE ENDS OF THE EARTH

Near the North and South Poles, the climate is so cold that ice exists all year round. Here we find ice caps. These are just like glaciers except that they cover whole continents, like Antarctica, or build up on the ocean surface as in the Arctic.

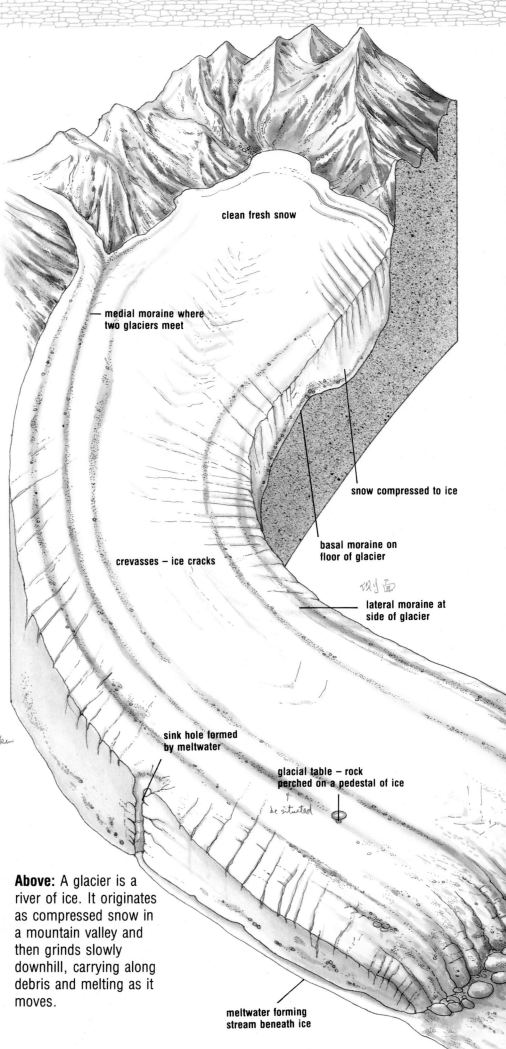

clean fresh snow

medial moraine where two glaciers meet

snow compressed to ice

basal moraine on floor of glacier

crevasses – ice cracks

lateral moraine at side of glacier

sink hole formed by meltwater

glacial table – rock perched on a pedestal of ice

meltwater forming stream beneath ice

Above: A glacier is a river of ice. It originates as compressed snow in a mountain valley and then grinds slowly downhill, carrying along debris and melting as it moves.

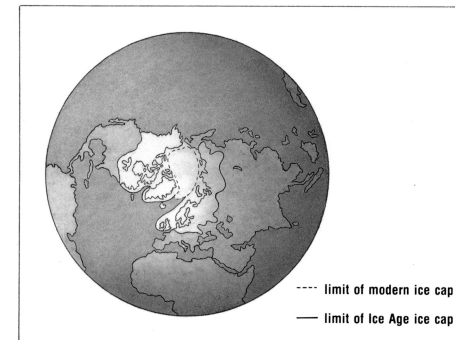

THE ICE AGE
During certain times in the past two million years, the ice caps were much larger than they are today. This was a result of the Ice Age when the climate of the world began to cool down. Ice caps spread out from the North and South Poles, and glaciers crept downward from the mountains. The chilled lowlands of Europe and North America had the climates and landscapes of today's Siberia.

- - - - limit of modern ice cap

——— limit of Ice Age ice cap

Right: The weight of a glacier erodes the bottom and the sides of its valley. When the glacier eventually melts, it leaves its valley with a distinct U-shape, as here in Switzerland.

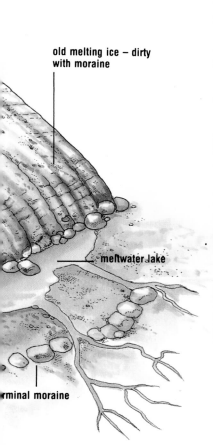

old melting ice – dirty with moraine

meltwater lake

rminal moraine

THE SEAS

More than two-thirds of the surface of our Earth is covered by the swirling mass of water that we call the oceans. So obvious is it, that it gives the planet a characteristic blue color. Indeed, when astronauts first saw the Earth from space, they called it The Blue Planet.

The water of the oceans is in constant movement. The winds traveling over the surface whip up the surface layers into a rippling motion that results in the waves. We eventually see these waves curling over and crashing onto the beaches.

As the rain pours down, some of it sinks deep into the ground where it dissolves away certain minerals in the rocks. These are washed down to the sea by rivers. Common salt is the most abundant of these dissolved minerals and gives seawater its salty taste. It stays in the ocean while pure water is evaporated from the surface and

becomes part of the water cycle once again.

OCEAN CURRENTS

Even more important are the great ocean currents that keep the waters moving around all the time. The Trade Winds blow constantly toward the equator, and these combine with the turning of the Earth to produce continuous westward-flowing equatorial ocean currents. The water thrust along the equator finally meets a continent and spreads north and south. Eventually it curls back and flows toward the equator again at the eastern side of the ocean.

This activity produces a huge swirl of water called a gyre, and each gyre usually occupies half an ocean. The parts of the gyre that flow away from the equator are warm currents. Those that flow toward the equator are cold, bringing along water from the colder regions.

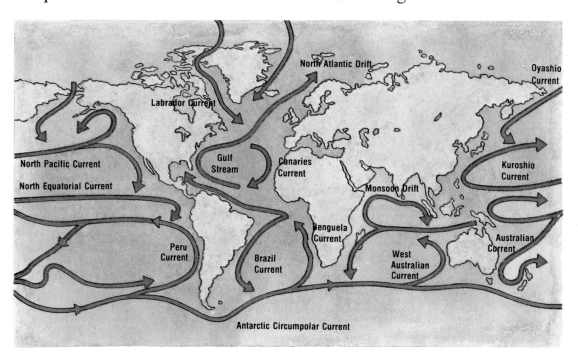

Ocean currents sweep around the world. They bring cold coastal waters and cool climatic conditions to hot countries, and warm up countries that would normally be very cold.

THE TIDES
The Earth is not alone in space. As we revolve around the Sun, so the Moon revolves around us. This movement of the Moon has an important effect on the oceans of our planet. The pull of the Moon's gravity forces the oceans into a bulge, and throws out another, matching, bulge on the opposite side of the Earth. It is this movement that causes the high and low tides that we see at the coasts.

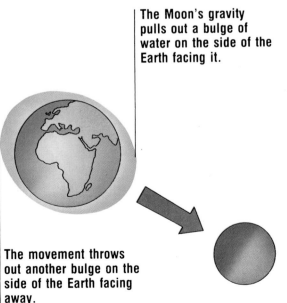

The Moon's gravity pulls out a bulge of water on the side of the Earth facing it.

The movement throws out another bulge on the side of the Earth facing away.

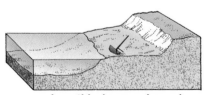

Low tide is experienced at each place between the bulges of water.

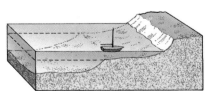

As the Earth turns beneath the pattern of bulges, each bulge produces a high tide.

SEASHORES

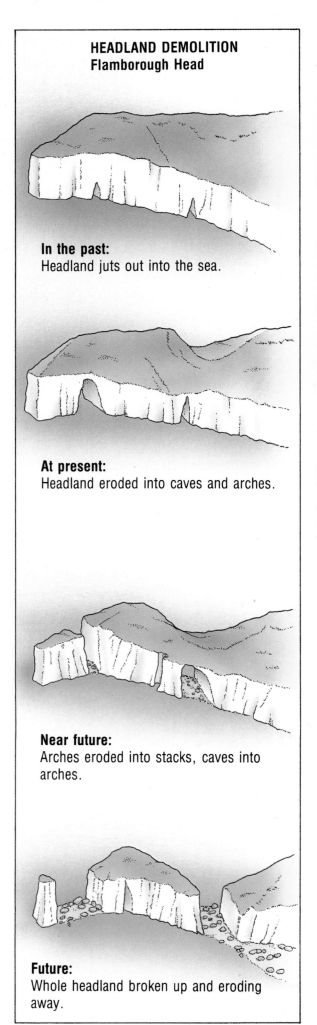

HEADLAND DEMOLITION
Flamborough Head

In the past:
Headland juts out into the sea.

At present:
Headland eroded into caves and arches.

Near future:
Arches eroded into stacks, caves into arches.

Future:
Whole headland broken up and eroding away.

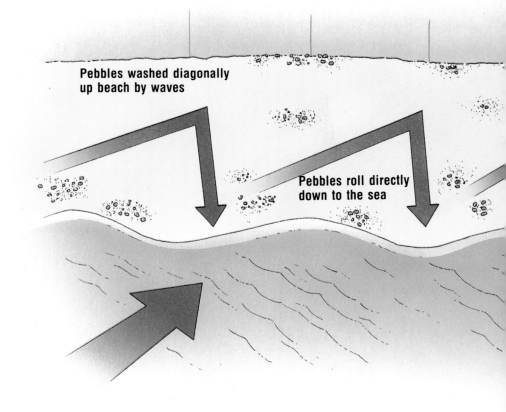

Pebbles washed diagonally up beach by waves

Pebbles roll directly down to the sea

The waves that are whipped up by the wind have a strong effect on the landscape of the seashores.

HEADLAND DEMOLITION

When waves approach a headland, they swirl around and smash into the headland from each side. Any land jutting out into the sea has its sides worn away, and soon becomes long and narrow. The waves pounding each side break open the rock along cracks that are already there, widening them and wearing them away to form sea caves.

Caves that are formed on each side of a headland may eventually meet in the middle and form a natural arch. As the erosion continues, the roof of the arch collapses, leaving the seaward part standing as an isolated sea stack. This is eventually destroyed by the relentless pounding of the waves.

BUILDING BEACHES

And what of the material worn away from the headlands? The rocks and sand are washed up on the shore to form pebbles and sandy beaches. Waves usually approach a beach at an angle. They wash the stone diagonally up the beach. As each wave subsides, it rushes straight back out again, carrying sand and stones with it. The next wave washes them farther on. As a result, pieces of sand and stone travel in a zigzag route along the beach. This movement is known as longshore drift.

Seaside resorts often lose their beaches because of this drift and fences called groines are built to try to stop the movement. The sand and stones then build up at one side of each groine giving the beach a saw-toothed pattern.

Above: Flamborough Head on England's east coast is being carved into caves and arches from each side.

Left: Waves strike a beach at an angle. Driven by prevailing winds, they move sand and pebbles along by longshore drift.

Shingle washed out from behind groine

Buildup of shingle against groine

WEATHERING AND EROSION

We can see the force of the weather in a storm. Rain lashes down in torrents, wind topples buildings, and lightning fells trees. Yet even when the rain is fine and dewy, and the wind is gentle and balmy, the weather still has a destructive effect on the landscape in general.

THE GENTLE RAIN
Rain soaks into the rocks, dissolving some of the minerals and weakening the strength of the rocks. Sometimes the rain seeps into pores and cracks, and then in cold weather the water freezes. Water expands when it freezes, and when this

happens, the ice forces the rocks apart along the cracks. Over a long period of time, a mountainside can disintegrate into long sweeping scree slopes of jagged boulders.

The weather gradually breaks down the surface rocks into a loose covering which, when mixed with dead vegetable matter, is called soil. Soil can be washed away by the rain, especially in dry areas where it does not rain very often. Any patch sheltered by an isolated boulder may remain after the surrounding soil is washed away. The boulder will then be left standing alone on a pedestal or a pillar of soil.

Below: Scree slopes consist of jagged boulders split away from exposed mountainsides. Ice expanding in crevices in a cliff face is responsible.

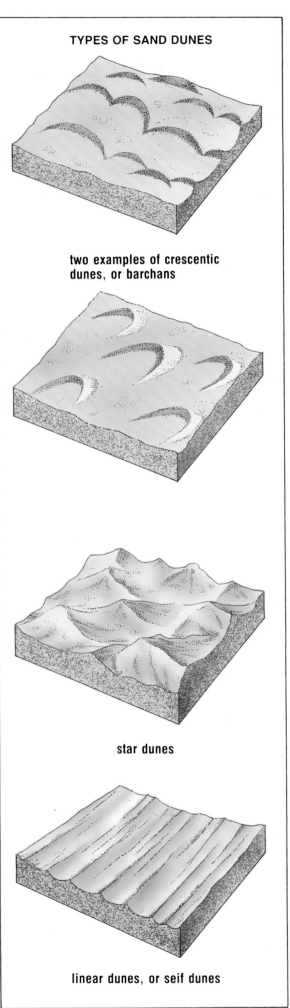

TYPES OF SAND DUNES

two examples of crescentic dunes, or barchans

star dunes

linear dunes, or seif dunes

WINDS

It is in hot, dry areas that the force of the wind really shapes the landscape. Particles of sand and dust are blasted against bare rock, polishing it into fantastic shapes. This usually happens close to the ground where the sand grains bounce along. As a result, even more sand and dust are produced and are blown over the surface, accumulating in heaps of sand called dunes. The dunes do not stand still. Sand particles are blown up one side and dropped down the other, so that the whole sea of dunes moves very, very slowly across the desert.

Above: The spectacular rocks of Bryce Canyon in Utah have been sculpted into shape by harsh sand particles blasted by the desert winds.

Right: Dunes are heaps of sand built up and blown along by the wind. Different wind patterns give rise to different types of dune.

53

ENERGY FROM THE EARTH

When you turn on an electric light, or ride in a car, you are using energy that has been collecting inside the Earth for hundreds of millions of years.

A plant consists largely of carbon, water and mineral substances. It uses the energy of the Sun to take the carbon from the carbon dioxide gas in the atmosphere, releasing oxygen as it does so. When the plant dies and decays, the oxygen of the atmosphere turns the carbon from the plant back into carbon dioxide and the energy is lost.

FOSSIL ENERGY

Sometimes, however, the plant is buried too quickly to allow the oxygen of the air to get at it. It does not decay away completely, and the carbon is trapped under the Earth. When this happens to large masses of plant material, it creates carbon-rich peat deposits. Over millions of years, peat may be compressed and turned into coal.

Nowadays we dig up the coal and burn it, allowing the oxygen of the air to reach it at last and turn it back into carbon dioxide. We harness the energy released to run our industries and heat our homes. In just a few hundred years, we have been using up the energy that it took millions of years to collect. At the same time, when we use fuel for our cars or to heat our homes, we are also returning to the atmosphere all the carbon dioxide broken down during those millions of years.

BLACK GOLD

What is true of coal is also true of oil. Oil is generated from the bodies of millions of sea creatures that collected on the sea bottom where there was no oxygen to decay them. The oil droplets, which eventually formed after the creatures were buried, seeped through the rock and floated up through the ground water. Sometimes they reached a layer of rock they could not seep through and collected there. The rocks below this cap rock became soaked with oil, producing an oil reservoir.

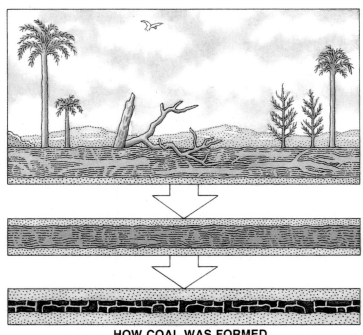

HOW COAL WAS FORMED

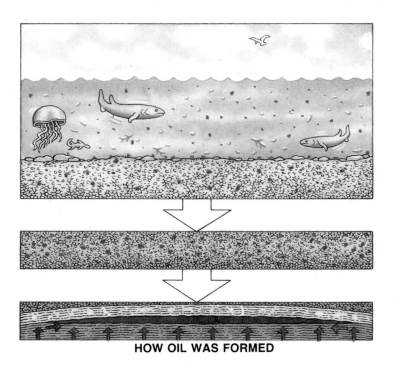

HOW OIL WAS FORMED

Top right: The increasing demand for oil forces drillers to extract it in the most inhospitable corners of the globe.

Right: Most of the world's major oil and coal fields are now well known. However, exploration companies are continually searching for new deposits.

54

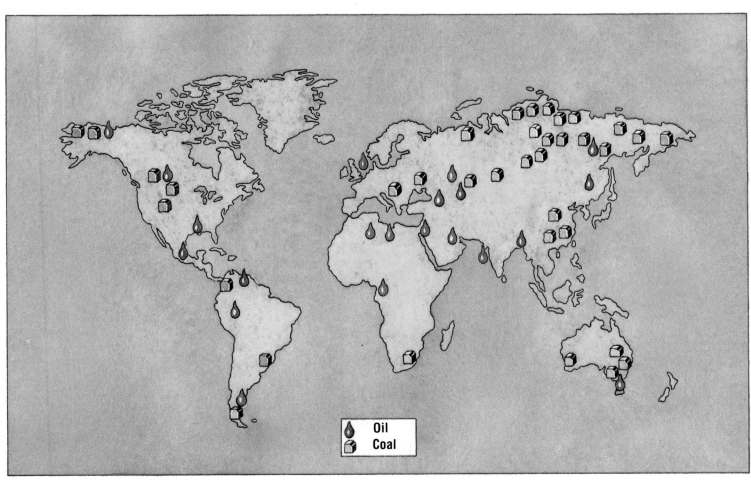

Oil
Coal

ALTERNATIVE ENERGY

We do not have to use up the energy trapped in the Earth by the buildup of dead things millions of years ago. Nuclear power, for example, relies on energy that was trapped in molecules when the Earth formed. However, this kind of energy is dangerous and difficult to handle.

A more attractive energy source is the Sun. Vast amounts of energy reach the Earth every second in the form of sunshine. This can be collected and turned into electricity directly by using solar batteries. There are also easier ways to use natural energy from our planet.

WIND POWER
The Sun heats different parts of the Earth at different rates. This is what causes the winds (see pages 58-59). Winds have been harnessed by windmills since the dawn of civilization. In the past they have been used to grind wheat and pump water. Now they can be used to turn generators to produce electricity. The winds also produce the waves.

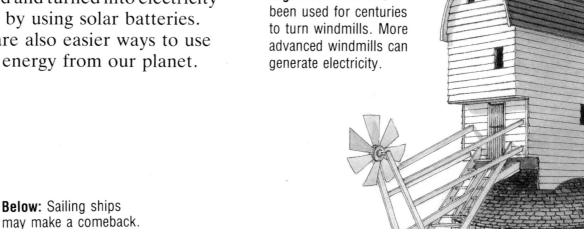

Right: The wind has been used for centuries to turn windmills. More advanced windmills can generate electricity.

Below: Sailing ships may make a comeback. Computer-controlled metal sails may some day be used to drive tankers.

WATER POWER

We can appreciate how much energy there is in the waves when we see storm damage. Methods for using energy from the waves have not, so far, been very successful, but further research continues.

The Sun also powers the water cycle, evaporating water from the oceans and allowing it to fall on the land and flow back to the sea as rivers. Since the beginning of history, people have been using water wheels to extract this energy. Water power has been used to grind wheat and to work the hammers of iron forges. Hydroelectricity is now produced by the same principle in such mountainous areas as northern Scotland and Norway.

LUNAR POWER

The pull of the Moon's gravity creates the tides (see pages 48-49). This energy, from the daily rise and fall of the water, is used in the estuaries of northern France to turn hydroelectric motors. This lunar (Moon) power is an energy source that may become more widely used in the future.

Above: The force generated by falling water can be harnessed in hydroelectric stations to generate electricity.

Below: Sunlight gathered by dozens of mirrors can be concentrated in a small area and used to heat a furnace.

Above: The daily sweep of the tide in and out of the mouth of the Rance River in northern France is used to generate electricity.

THE ATMOSPHERE

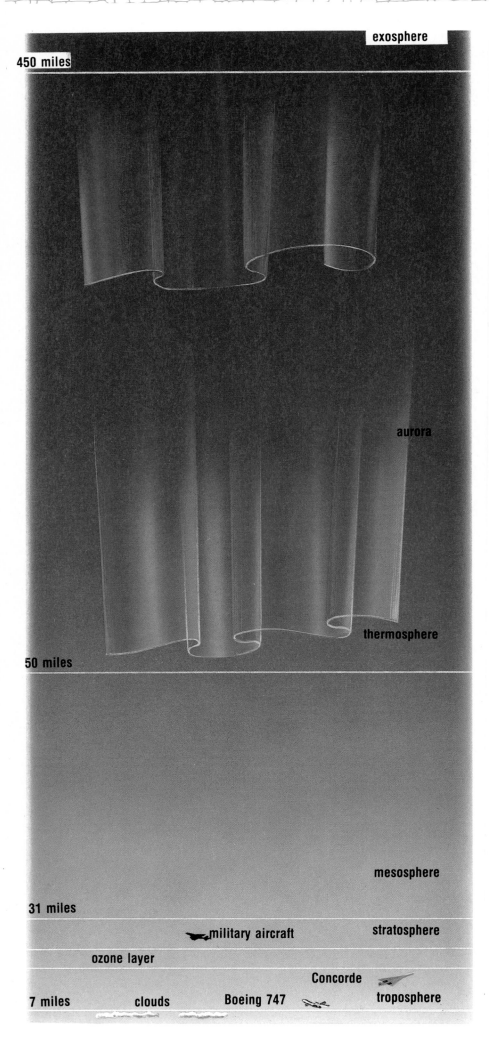

exosphere

450 miles

aurora

thermosphere

50 miles

mesosphere

31 miles

military aircraft

stratosphere

ozone layer

Concorde

7 miles   clouds   Boeing 747   troposphere

When we looked at how the Earth was built of different layers, becoming lighter from the core outward, we did not describe the lightest layer of all. This is a thin envelope of gas that surrounds the entire planet. We call this envelope the atmosphere – from the Greek *atmos* meaning vapor.

The atmosphere is a mixture of gases, most of which is nitrogen. The nitrogen, however, is relatively unimportant compared with the second most abundant gas, oxygen. Without oxygen, we could not live.

SUN-DRIVEN WINDS
The Sun heats the atmosphere more at the equator than elsewhere because it is directly overhead. When air becomes warm, it tends to rise, and cooler air rushes in to take its place. As a result, there are winds that blow constantly toward the equator. These are called the Trade Winds. They do not blow directly north and south, but are deflected to the west by the turning of the Earth.

To the north and south of the equator, the air cools and descends once more. When it hits the ground, it spreads toward the equator, becoming the Trade Winds again, and toward the cooler regions where it becomes the warm Westerlies. Like the Trade Winds, these are deflected from a true north-south course by the turning of the Earth.

Over the North and South Poles the cold air descends and cold winds blow outward. These cold winds meet the Westerlies head-on and cause the unsettled climate conditions in the temperate latitudes (see pages 68-69).

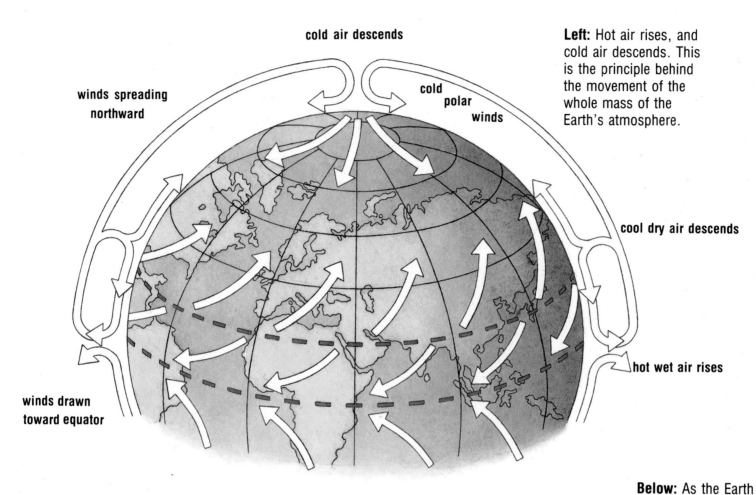

cold air descends

winds spreading
northward

cold
polar
winds

Left: Hot air rises, and cold air descends. This is the principle behind the movement of the whole mass of the Earth's atmosphere.

cool dry air descends

hot wet air rises

winds drawn
toward equator

Below: As the Earth turns, the winds are twisted east and west. The climate patterns of the world arise from this general circulation pattern.

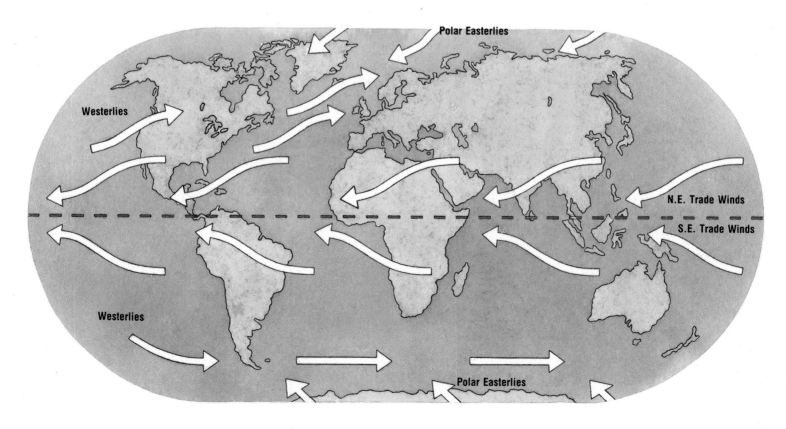

Polar Easterlies

Westerlies

N.E. Trade Winds

S.E. Trade Winds

Westerlies

Polar Easterlies

WEATHER AND SEASONS

What is the difference between climate and weather? Climate is what we are supposed to get, while weather is what we do get! Climate is the average set of atmospheric conditions – rainfall, temperature, wind – in a particular place over a long period. Weather is the daily change of that pattern.

WIND AND RAIN
The broad climate pattern is established by the way that the winds travel over the globe. We get rainy climates where wet winds drop their moisture, as when the air rises over mountains or over masses of colder denser air. We get dry climates where only dry winds blow. The shapes of the continents and the mountain ranges complicate the worldwide pattern.

The daily variations in this pattern give us the weather. However, there are also variations in the climate between one part of the year and another. These variations produce the seasons.

Above: Desert conditions represent an extreme type of climate. Where no moisture reaches an area, no plants can grow and no animals live.

Right: Monsoon conditions represent the other extreme, with torrential rain at one time of the year. However, the monsoon is seasonal, and the rest of the year is dry.

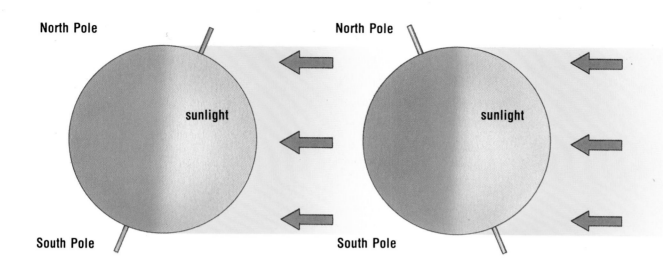

North Pole

sunlight

South Pole

North Pole

sunlight

South Pole

Above: Northern summer, and the North Pole tilts toward the Sun, exposing northern areas to more sunlight.

Above: In the northern winter, the North Pole tilts away, and the southern hemisphere receives more sunlight.

THE YEARLY CHANGE

The Earth's axis – the imaginary line around which it spins – is not vertical. It leans toward the Sun at one part of the year, and away from the Sun at another. This means that each place on the Earth's surface will receive a different amount of sunlight at different times of the year, and so there are different seasons.

In temperate regions, where it is not too hot and not too cold, there are four distinct seasons: spring, summer, fall, winter. When it is summer in the Northern Hemisphere, it is winter in the Southern. Nearer the Poles there are two seasons – summer and winter. The difference between the two is very great and there is hardly time for a spring or fall between them. In tropical regions there may also be two seasons – dry or wet. Monsoon areas have three – a hot season, a cool season and a rainy season, which can cause terrible floods.

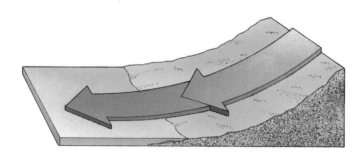

Cool season
The wind blows from inland Asia, from the Himalaya Mountains, out toward the Indian Ocean.

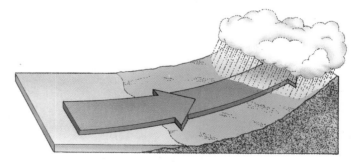

Rainy season
Warm air rises over the Asian land mass, drawing in wet winds from the Indian Ocean and dropping rain.

RAINFORESTS

toucans

Along the steamy equator, the Sun is almost overhead for most of the year. This heats up the land and warms the air above it. The hot air rises and air from the north and the south moves in to take its place. Usually the winds formed this way blow from the sea, bringing moisture with them. In the rising air stream, the moisture condenses and drops as torrential rain almost every day. These hot and wet conditions are ideal for plant growth and tropical rainforests abound.

A HOTHOUSE ENVIRONMENT

So many different kinds of plants can live under these hothouse conditions that hundreds of different species live within a small patch of ground. The tallest trees, called the emergents, can be 100 feet high and stick up above the rest of the forest.

Below them is the canopy – a rooftop of entwined branches formed as all the trees struggle upward to reach the Sun. The boughs are covered in creepers that grow upon other plants. Down below is darkness, all the sunlight blocked out by the leaves and branches of the canopy. Very little grows here.

CONSTANT GROWTH

The largest rivers of the world – the Amazon, the Zaire, the Mekong – flow through these areas, fed by the constant rains. Along their banks the forest canopy sweeps down to the water's edge, catching the sunlight that comes in from the side.

Wherever an old tree falls, the light floods into the clearing and a thick growth forms on the forest floor. Eventually the gap is plugged by new trees.

With so many kinds of plants living in one area there are many different kinds of food. As a result, many different animals live here, too. Toucans and monkeys live in the canopy; sloths and iguanas on tree trunks; and peccaries and okapis on the forest floor. If the rain forests were destroyed by careless farming, we would lose lots of our plant and animal life.

Below: The tropical rainforests are found along the equator, where the Trade Winds bring moist air all year. The world's biggest rivers are found here.

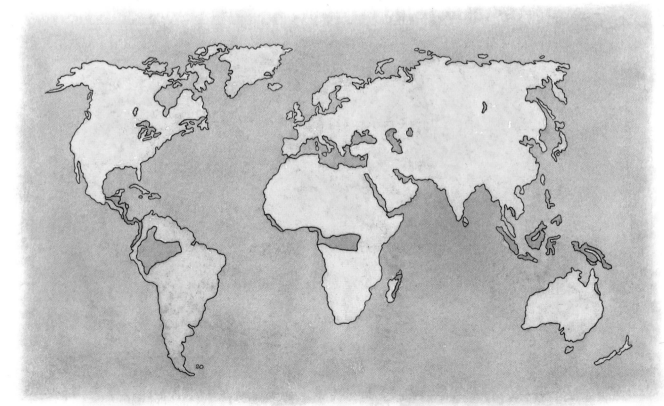

monkey

sloth

iguana

okapi

Above: Thick jungles clothe the slopes of the mountains in tropical rainforest areas, such as here, in Costa Rica.

Left: So much grows in the rainforest that there is food to be found everywhere, from the forest canopy to the gloomy floor. Animals live at all forest levels.

63

GRASSLANDS

To the north and south of the equatorial rainforest belt lie regions of tropical grassland. The rains fall here only at certain times of the year. This is usually in the summer, when the sun is directly overhead and the moist Trade Winds blow into the region. During the rest of the year, these regions come under the influence of the hot, dry conditions that produce the deserts farther to the north and south.

WIDE-OPEN SPACES
Trees do not do well under these conditions, but grasses do. As a result, there are belts of grasslands to the north and south of the equatorial rainforest belt. Perhaps the most famous are in Africa, both in the north and the south.

In the Southern Hemisphere, grasslands are found as the pampas of South America and the outback of Australia. The Northern Hemisphere grasslands are the prairies of North America and the steppes of Asia. These, however, are temperate grasslands and are often made up of vast areas of gently rolling plains. They occur because they are a long way from the sea and receive little rainfall. However, they do have hot summers and very cold winters.

Grass thrives because it can regrow from underground stems and roots. In the dry season, all the leaves and stalks can be burned away by bush fires, and throughout the year grazing animals will eat them down to the ground. Under the same conditions, most bushes and trees would be destroyed, but the grasses can grow again from under the ground.

zebra

antelope

kangaroo

ANIMALS OF THE PLAINS
The grassland supports a particular type of animal. You can usually tell a grass-eating animal – a grazer – by its shape. It will have long running legs to escape from danger. There is nowhere to hide from predators on the grassland. It will have strong jaws, because grass is very tough to chew. It will have a long face, so that when it is eating the grass, its eyes can still keep a lookout. Zebras, antelope and even kangaroos are typical grazing animals.

Above: Grassland animals are fast, and can run or hop quickly over the open plains away from hunting animals. Long jaws contain strong teeth for chewing the grass.

Above: Hunting animals of the grasslands, such as this leopard, must be camouflaged against the grass. They must also be able to produce a burst of speed for the attack.

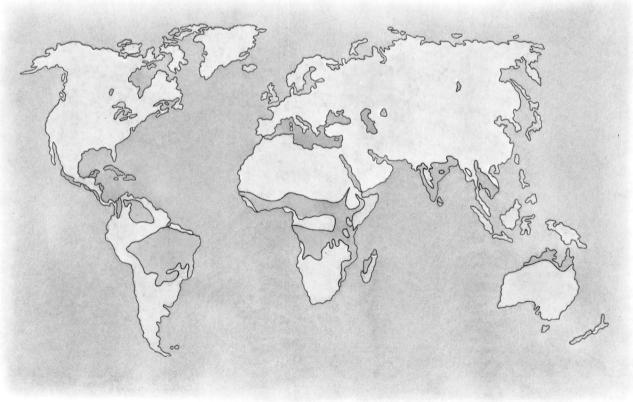

Right: Most grasslands lie between the rainforest belt and tropical deserts. The climate alternates between the two extremes.

65

DESERTS

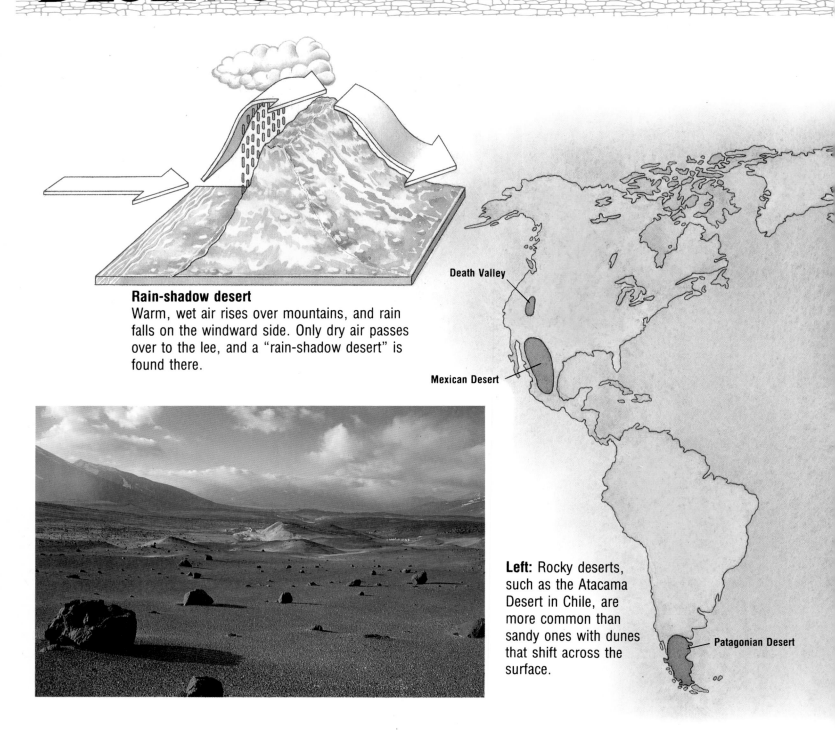

Rain-shadow desert
Warm, wet air rises over mountains, and rain falls on the windward side. Only dry air passes over to the lee, and a "rain-shadow desert" is found there.

Death Valley

Mexican Desert

Left: Rocky deserts, such as the Atacama Desert in Chile, are more common than sandy ones with dunes that shift across the surface.

Patagonian Desert

At the equator the hot air rises, dropping its moisture into the tropical rainforest below. High in the atmosphere, this air spreads north and south and begins to cool. As it cools it descends, and this dry air reaches the ground at about the latitudes of the tropics of Cancer and Capricorn. As a result the climate here is very dry, nothing grows, and the landscape is one of desert.

Along the northern belt, there are the deserts of Mexico, the Sahara and Arabia. Along the southern belt are the Kalahari of Africa, and the Gibson and Simpson Deserts of Australia.

The heat that is absorbed during the day is lost quickly again at night, because there is no blanket of cloud to hold in the warmth. As there are no plant roots to hold the soil together, it breaks down easily to sand, and the wind blows it away. Most desert areas are bare rock and stony rubble. Only about a fifth are made from sand.

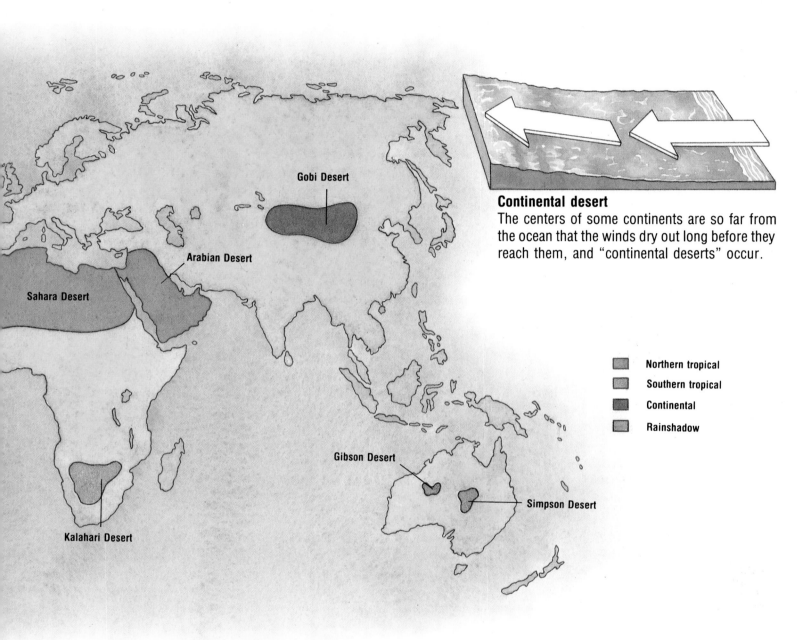

Continental desert
The centers of some continents are so far from the ocean that the winds dry out long before they reach them, and "continental deserts" occur.

Northern tropical
Southern tropical
Continental
Rainshadow

Gobi Desert
Arabian Desert
Sahara Desert
Kalahari Desert
Gibson Desert
Simpson Desert

DRY WINDS
Other types of desert are found in the lee, or sheltered side, of mountain ranges. Winds from the sea always drop their rain on the windward sides of mountains, as the air rises. Only dry winds pass over the mountains to the rainshadow on the other side. Such rainshadow deserts include the Patagonian Desert of South America and Death Valley in North America.

Continental deserts lie in the hearts of great continents, so far from the sea that no moisture reaches them, whichever way the wind blows. The most famous is the Gobi Desert in central Asia.

DESERT ANIMALS
Only very special animals can live in the desert. The creatures that live there do not drink. Plant-eating animals obtain all their water from the few plants that manage to survive. Meat-eating animals get theirs from the flesh of these plant-eaters. Most animals burrow deep into the soil to hide away from the Sun, and only come out at night.

Above: The desert areas of the world lie mainly along the tropics, in the rainshadows of the main mountain ranges, and in the centers of continents.

67

TEMPERATE ZONES

Above: Temperate regions have warm summers and cool winters. The trees are usually deciduous, losing their leaves and becoming dormant during winter.

Right: The variable weather experienced in temperate zones is due to the constant conflict between warm winds from the tropics and cold winds from the Poles.

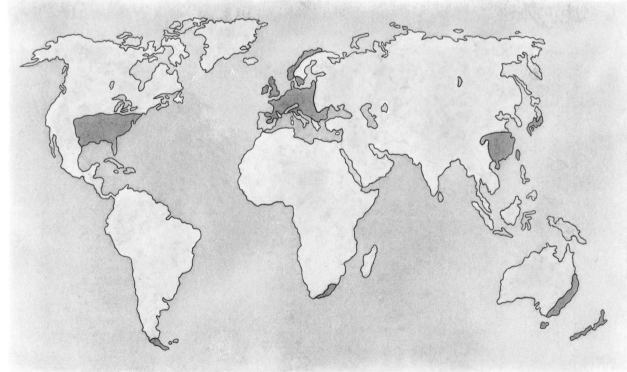

A temperate climate is one that is not too hot and not too cold, not too wet and not too dry. Most of Europe, the coastal states of North America, the eastern seaboard of Asia, and the southern tips of South America, Africa and Australia, can be thought of as lying in the temperate zones. The presence of the sea helps to keep some areas temperate, cooling regions that should be hot, and warming others that should be cold.

BATTLING FRONTS

In the Northern Hemisphere most temperate climates lie along a belt where warm winds from the southwest meet cold northeasterly winds from the Poles. When different air masses like these meet, the boundary between them is called a front. The weather along a front tends to be unstable.

SEASONS

Seasonal changes are very noticeable in temperate climates. Temperatures are higher and days are longer in summer than in winter. The Sun appears higher in the sky, so it produces warmer weather.

In the winter everything is cold and nothing grows. This is when animals, such as squirrels and mice, hibernate with their stores of fruits and nuts until the weather becomes warmer again. Come the spring, the shoots push through the ground, and the buds appear on the trees. In the summer the plants put on their strongest growth. Finally in the fall the fruits and seeds are ready to be shed. The leaves, having finished their work of replenishing the tree's energy, fall off. Then it is winter again.

BATTLING FRONTS

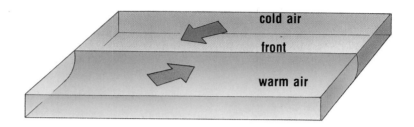
cold air
front
warm air

Air masses turn toward each other.

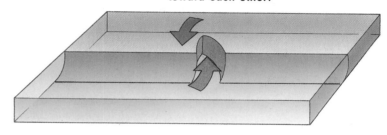

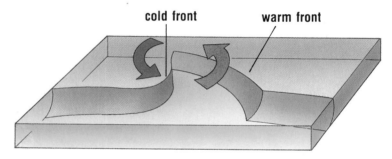
cold front   warm front

Above: Frontal systems arise as cold polar air slides past warm tropical air. The two air masses spiral around one another, producing very changeable weather.

Left: More predictable temperate weather is found in Mediterranean regions, where polar air has little influence and tropical winds bring rain in winter time.

CONIFEROUS FORESTS

In the northern parts of the temperate zone, the winters become very harsh indeed. Water exists only as ice for most of the year and cannot be used by animals or plants. This is the region of the coniferous forests – the largest area of unbroken forest in the world – stretching from Europe, across Asia, and across Canada, too. In the Southern Hemisphere there are no large areas of land this close to the Pole, and so there is no large area of coniferous forest here.

COLD-WEATHER TREES

A conifer tree can put up with the dreadful conditions found in the cold latitudes. Its conical shape can shake off the masses of snow that would topple any other shape of tree. The leaves grow in the form of needles and have a waxy covering so that they do not lose too much water in the summer. The roots spread out over a wide area close to the surface to hold the tree in the thin soil. Coniferous forests, unlike tropical rainforests, have few species of trees. There are not many

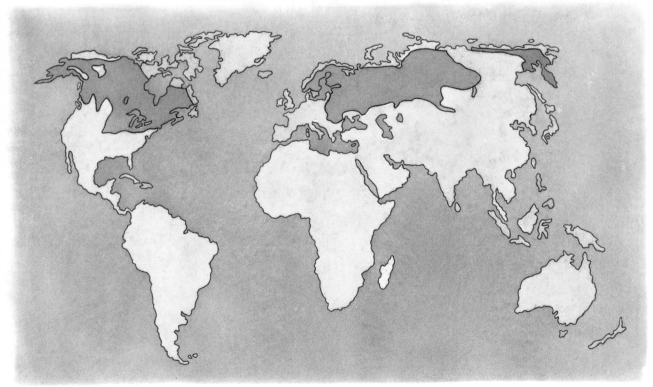

Above: Coniferous trees are evergreen. The leaves are always on the trees, ready to make use of sunlight as soon as it appears.

Left: The northern coniferous forest is the biggest uninterrupted forest area in the world. Smaller forests grow in mountainous areas.

that can withstand these cold conditions.

The conditions that produce coniferous forests are found toward the Poles. They are also found toward the tops of mountains, and mountain regions have their zones of coniferous forests as well. These are called montane forests, as opposed to the boreal forests or the taiga found in the cold latitudes.

COLD-WEATHER ANIMALS

Animals that live in coniferous forests are basically cold-weather animals. They do most of their eating in the short spring and summer, when new shoots are growing. Large animals, like moose, build up layers of fat that nourish them during the winter when there is no food growing.

Small animals, like squirrels, hide supplies of seeds and other food to feed them during the cold, dark winter months. Only a few birds, like grouse, can actually eat conifer needles and find some nourishment in them.

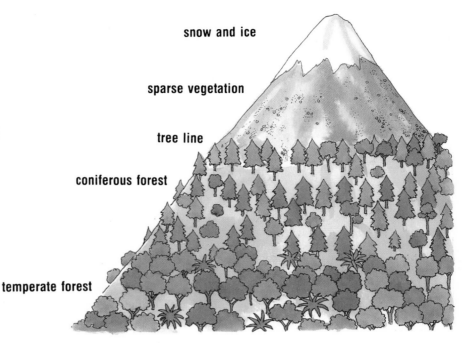

snow and ice

sparse vegetation

tree line

coniferous forest

temperate forest

As you climb a mountain, the vegetation changes the way it changes as we approach the Pole. Coniferous forest lies above the temperate regions and below the icy peaks.

Below: The most successful coniferous forest animals are those that can eat the conifer needles, or store away the seeds for use during the winter.

moose

grouse

squirrel

chipmunk

71

POLAR REGIONS

At the far north and the far south of the globe, the Sun is always low in the sky. In the summer it seems only to be peeping up over the edge of the world. In the winter it does not rise at all. At the Poles themselves, the year is divided into six months of permanent light and six months of permanent darkness.

THE ICY ENDS OF THE EARTH

The poles are covered with ice. At the north there is the Arctic Ocean with a floating ice cap, and a land-based ice cap over Greenland. At the south there is the continent of Antarctica, covered by a sheet of ice thousands of yards thick. Both produce icebergs. The icebergs of the north are spiky and irregular, having broken away from the Greenland glaciers. Those of the Southern Ocean are huge and flat. They have broken free from the ice sheet that has spread outward from Antarctica and into the ocean.

TUNDRA LIFE

Away from the poles there are areas that are a little more habitable. In these areas the covering of snow and ice melts in the summer. However, the soil beneath remains frozen – a condition known as permafrost – and the water cannot drain away. The summer landscape becomes one of lakes and bogs. Tough vegetation lives here, such as hardy grasses, mosses and lichens. The trees are limited to a few stunted birches. The environment is known as the tundra in northern Europe and Asia, and the muskeg in the north of Canada.

In the summer herds of animals such as reindeer migrate northward into the tundra to feed on this growth, having wintered in the coniferous taiga. Insects like flies and mosquitoes suddenly become active in and around the lakes and bogs. Birds like larks and martins flock to the area to feed on these insects during the short, bleak summer months.

The North Pole
The northern ice cap floats on the Arctic Ocean, with only a few islands poking through.

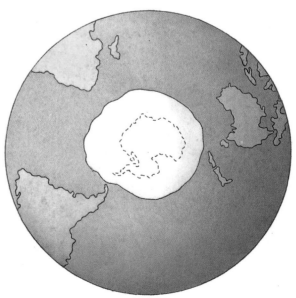

The South Pole
The southern ice cap covers the continent of Antarctica, and spreads out into the oceans round about.

72

Above: Around the fringes of the ice caps lies the tundra – a region of permanently frozen subsoil supporting a sparse growth of low plants and stunted trees.

Left: Ice caps creep slowly out into the ocean. Chunks break from the end and form icebergs that can be a danger to shipping.

THIS FRAGILE EARTH

Our world is dirty. Over the past few hundred years, we have made it so. As our civilization has developed, our way of life has improved. No longer do we have to gather berries and pick nuts for our food. Nor do we have to hunt the wild animals for their meat, and for their skins to use as clothing. Instead we can grow our food on farms, and make our clothing and the other necessities of life in factories. Life is easier than it was for our ancestors. Unfortunately there are drawbacks.

CHANGING THE LAND

When we clear away forest or grassland to make the fields for our farms, we are destroying the natural habitats that we have just described (see pages 62-72). If we grow unsuitable plants, the soil breaks down and washes away, and then nothing can grow.

Our factories need minerals for raw material, and fuel for power. These have to be obtained from quarries and mines that eat into the natural rocks and countryside.

Above: Industry causes waste and pollution. Unchecked, it can scar the landscape, such as here in an abandoned tin mine in England.

Right: To clear enough space for farming, we have always had to chop down the natural vegetation. In the rainforests, this is happening on a disastrous scale.

POLLUTION

Every industry has waste products of one kind or another. If these wastes are liquid, they may be poured into rivers and seas, and poison them. If they are solid, they may be dumped in garbage dumps that clutter up the landscape. When fuel is burned, it gives off carbon dioxide and other gases. These alter the mixture of gases in the atmosphere, which may change the climates all over the world – the so-called greenhouse effect. Other gases, such as CFCs, can destroy the ozone layer – a layer of ozone gas in the atmosphere that protects us from the harmful rays of the Sun.

All this represents a change to the Earth's environment. Throughout the planet's history the environment has been changing. However, the changes have never been so fast as those that we are causing now. We must make sure that we know what these changes are so that we can control them for the long-term good of our planet.

Below: The Earth is a jewel in space. There is only one, and it must be treasured.

INDEX

Figures in **bold** refer to captions.

A

Africa
 Great Rift Valley 19, **19**, 35
 Sahara Desert 66
 southern 69
 Zaire River 62
America *see* North America *and* South America
amphibians, early 40
Andes 35
andesite 20, **21**, 26, 27
animal life
 of coniferous forests 71, **71**
 of deserts 67
 early 40-1, **41**
 of grasslands 64, **64**, **65**
 of polar regions 72
 of tropical rainforests 62, **63**
Antarctica 46, 72, **72**
anticline 32, **32**
Archaean Era 36
Arctic Ocean 46, 72, 72
Asia
 climate **61**, 69
 coniferous forests 70
 eastern seaboard 69
 Gobi Desert 67
 Himalaya Mountains 19, 35
 Mekong River 62
 steppes 64
 tundra 72
 Ural Mountains 35
Atlantic Ocean, formation of 19
atmosphere, Earth's **13**, 42, 58, **59**
 formation of 10
 pollution of 75
 in prehistoric times 40
aurora borealis 15
Australia
 deserts 66
 outback 64
 southern 69
axis, Earth's, angle of 61

B

basalt 20, **20**, 26, 27
beaches 38, **39**, 51, **51**
beds, sedimentary rock 28, 38
birds
 of coniferous forests 71
 early 41
 of polar regions 72
 of rainforests 62
block mountains 35, **35**
breccia 28

C

Cambrian Period 36
carbon 54
carbon dioxide **43**, 54
caves
 in limestone areas **43**
 sea 51, **51**
Cenozoic Era 36, 41
CFCs 75
climate 48, 58, **59**, 60, **60**, 61

in deserts **60**, 66-7
 effect of pollution on 75
 in grasslands 64, **65**
 during Ice Age **47**
 in monsoon regions **60**, 61
 in polar regions 61, 72
 temperate 58, 61, **68**, 69, **69**, 70
 tropical 61, 62, **62**, 64
coal 54, **54**
Coast Ranges 35
conglomerate 28
conservation 74-5, **75**
continental crust 13, 16, 19
continents, formation of **18**, 19, **19**
core, Earth's 13, **13**, 15
crust, Earth's
 continents **18**, 18, **19**
 earthquakes 22, **22**, **23**, 33
 folds and faults 22, 32, **32**, 33, **33**
 mountains **18**, 19, **34**, 35, **35**
 movement of 16, **16**, **17**, **18**, 19, **19**
 substance of 13, **13**; *see also* rocks
 volcanoes 10, 16, 18, 19, 20, **20**, **21**
currents
 ocean 38, 48, **48**
 river 38, **39**

D

deserts **60**, 66, **66**, 67, **67**
dinosaurs 41
dolerite 27
drift, longshore 51, **51**
dunes 38, **38**, 53, **53**

E

Earth
 atmosphere *see* atmosphere, Earth's
 climate *see* climate, Earth's
 conservation 74-5, **75**
 energy from 54-7
 formation of 10
 geographical features 44-53
 geological history 36, 38, **38**
 life on *see* animal life *and* plant life
 magnetic field 15, **15**
 natural resources *see* energy, minerals, trees *and* water
 shape **13**
 structure 13, **13**
 substance 13, **13**; *see also* rocks
 surface *see* crust, Earth's
earthquakes 22, **22**, **23**, 33
Earth Science 9
electricity *see* energy
energy
 fossil 54
 lunar (from Moon) 57, **57**
 nuclear 56
 solar (from Sun) 56, 57, **57**
 from water 57
 from wind 56, **56**

environment
 changes to 75
 see also climate *and* Earth
equator 48, 58, 62, **62**, 66
erosion 44, **44**, 45, **45**, **50**, 51, 52
Europe
 coniferous forests 70
 temperate climate 69
 tundra 72
evolution 40-1

F

faults 22, 33, **33**
 see also block mountains
fish, early 40
fold mountains **34**, 35
folds 32, **32**
 see also fold mountains
forests
 boreal 71
 coniferous 70, **70**, 71, **71**
 montane **70**, 71, **71**
 tropical rain 62, **62**, **63**, **74**
fossil energy 54
fossils 9, 36, 40, **40**
fronts 69, **69**
fuels, fossil 54

G

gabbro 27
geographical features of Earth 44-53
 see also continents, faults, folds, icebergs, ice caps, mountains, oceans, rivers, valleys *and* volcanoes
geological time 36
geology 9
glaciers 46, **46**, **47**
gneiss 31
grabens 33, **33**
granite **24**, 27, **27**
grasslands 64, **64**, **65**
Great Rift Valley 19, **19**, 35
greenhouse effect 75
ground water 42, **42**, **43**, 44
groines 51
gyres 48

H

habitats, natural *see* animal life *and* plant life
 destruction of 74
Himalayas 19, 35
horsts 33, **33**
humans, first 41
hydroelectricity 57, **57**

I

ice 46, **46**, 52, **52**
Ice Age **47**
icebergs 72, **73**
ice caps 46, **47**, 72, **73**
igneous rocks 26, **26**, 27, **27**, 31
island chains, formation of 16, 19

L

landscape, Earth's 44-53
 see also continents, faults, folds, icebergs, ice caps, mountains, oceans, rivers, valleys *and* volcanoes
lava 20, **20**, **21**, 26, 27
limestone 28, 38, **43**
longshore drift 51, **51**
lunar energy 57, **57**

M

magma 20, 27
magnetic field, Earth's 15, **15**
mantle, Earth's 13, 16, **18**
marble **31**
Mesozoic Era 36, 41
metamorphic rocks **24**, 30, **30**, 31, **31**
meteorites **11**
mammals
 early 41
 of coniferous forests 71
 of grasslands 64
 of polar regions 72
 of rainforests 62
minerals **24**, **24**, 48
 ore **24**, 25
 silicate 25
Modified Mercalli scale 23
monsoon **60**, 61
Moon
 effect on tides **48**, **49**
 energy from 57, **57**
moraine 46
mountains
 block 35, **35**
 fold **34**, 35
 forestation of **70**, 71, **71**
 formation of **18**, 19, **34**, 35
mudstone 28
muskeg 72

N

natural resources *see* energy, minerals, trees *and* water
 depletion of 54, 62, 74, **74**
North America
 coastal states 69
 Coast Range Mountains 35
 coniferous forests 70
 deserts 66, 67
 prairies 64
 Rocky Mountains 35
Northern Lights **15**
nuclear power 56

O

Ocean
 Arctic 72, **72**
 Atlantic 19
 Pacific 16, **20**
ocean currents 38, 48, **48**
oceanic crust 13, 16, 19
oceans 42, **42**, 48
 currents 48, **48**
 formation of **18**, 19
 tides **48**, **49**

oil 54, **54**
ores
  iron **24**, 25
  native **24**
outback, Australian 64
oxygen 54, 58
ozone layer 75

P
Pacific Ocean, islands of 16, **20**
paleogeography 39
Paleozoic Era 36, 40
pampas, South American 64
permafrost 72
Phanerozoic time 36
plains 64, **64**
planets, formation of 10
plant life
  coniferous 70, **70**, 71, **71**
  early 40
  of grasslands 64
  of polar regions 72, **73**
  of temperate regions **68**
  of tropical rainforests 62, **63**
plate margins **18**, 22, 30
plate movements 16, **17**, **18**, 19,
  20, 22, **22**, 30, 32, **34**, 35
plate tectonics 16
Poles, North and South
  geographic 15, **15**, 46, **47**, 58,
    61, **61**, 72
  magnetic 15, **15**
pollution **74**, 75
power
  hydroelectric 57
  lunar 57, **57**
  nuclear 56
  solar 56, 57, **57**
  water 57
  wind 56, **56**
prairies, North American 64
Precambrian Era 36
Proterozoic Era 36, 40

R
Rain
  effect on landscape **43**, 44,
    48, 52
  as factor in climate 60, **60**,
    61, **61**, 62, 64, 67
  as part of water cycle 42,
    42, 57
rainforests, tropical 62, **62**, **63**
  destruction of 62, **74**
rainshadow deserts **66**, 67, **67**
reptiles
  early 40-1
  of rainforests 62
Richter scale 23
ridges, oceanic 16, **16**
rift valleys **18**
  Great Rift Valley 19, **19**, 35
river currents 38, **39**
rivers 42, **42**, 44, **44**, **45**
  largest in world 62, **62**
Rockies 35
rocks 24-35
  igneous 26, **26**, 27, **27**, 31
  metamorphic 24, 30, **30**, 31, **31**

minerals making up 24, **24**
  sedimentary 28, **28**, **29**, 38
  study of 38, 39

S
salt 48
sand dunes 38, **38**, 53, **53**
sandstone 28, **29**, 38
saturation zone 42, **42**
schist 31
scree slopes 52, **52**
sea caves 51, **51**
seas *see* oceans
seashore 38, **39**, 51, **51**
seasons 60, 61, 69
sea stacks 51
sedimentary rocks 28, **28**, **29**, 38
shale 28, 38
silica 25, 27
slate **30**, 31
snow 46, **46**
soil 52
solar energy 56, 57, **57**
Solar System, formation of 10, **11**
South America
  Amazon River 62
  Andes Mountains 35
  pampas 64
  Patagonian Desert 67
  southern 69
springs 42, **42**, 44
stacks, sea 51
steppes, Asian 64
  stratigraphy 39
sulphates 25
Sun
  effect of on Earth's atmosphere
    58, 61, 62
  effect of Earth's atmosphere on
    radiation from 75
  energy from 56, 57, **57**
  formation of 10
supercontinents 19
syncline 32, **32**

T
taiga 71, 72
tectonics, plate 16
temperate climates 58, 61, **68**,
  69, **69**, 70
tides **48**, 49
  energy from 57, **57**
tors 27
Trade Winds 48, 58, **62**, 64
Trees
  coniferous 70, **70**, 71, **71**
  of polar regions 72, **73**
  of rainforests 62, **63**, **74**
trilobites **40**
tropical grasslands 64
tropical rainforests 62, **62**,
  63, **74**
tundra 72, **73**

U
Ural mountains 35

V
Valleys
  rift **18**, 19, **19**, 35
  U-shaped 46, **47**
vegetation *see* plant life
volcanoes 10, 16, **18**, 19, 20,
  21, 26

W
water
  ground 42, **42**, **43**, 44
  river 44-5
  sea 48
water cycle 42, **42**, 57

water power 57, **57**
water table 42, **42**
waves 48
  effect on landscape 51, **51**
  energy from 57
weather 60
Westerlies 58
wind
  effect on climate **59**, 60, **61**, 62,
    67, **68**, 69
  effect on landscape 51, **51**,
    53, **53**
windmills 56, **56**
Winds, Trade 48, 58, **62**, 64

ACKNOWLEDGMENTS

Original synopsis by Neil Curtis

PHOTOGRAPHS

Ardea: 19 (Richard Waller)
Bruce Coleman Ltd.: 15 (Jane Burton)
Dougal Dixon: 33, 63, 65
Geoscience Features Picture Library: 24 (M. Hobbs), 27, 38 (Dr. B. Booth), 39 (top), 47, 50-51 (Dr. B. Booth), 52, 73 (top)
Nature Photographers Library: 30 (Derick Bonsall), 48-49 (Paul Sterry)
RIDA Photo Library: 29, 32 and 39 (bottom) (Richard Moody), 70 and 74 (top) (David Bayliss)
Science Photo Library: 11 (Baker/Milon); 13 (NASA); 14 (Jack Finch); 17 (Simon Fraser); 20 (Soames Summerhays); 21 (Stewart Lowther); 22-23 (David Leah); 24bl (Roberto de Gugliemo), 24bc (Sinclair Stammers); 25t (Sinclair Stammers); 25b (Roberto de Gugliemo); 31; 40 (Sinclair Stammers); 43 (David Campione); 44 (Harry Nor-Hansen); 45b (Dr. Robert Spicer); 45t (John Heseltine); 53 (David Parker); 55 (Harry Nor-Hansen); 60 (Sinclair Stammers); 66 (Simon Fraser); 68 (John Heseltine); 69 (John Heseltine); 73b (Simon Fraser); 74b (Dr Morley Read); 75 (NASA)
Stockphotos: 34 (Gordon L. Kallio)

ILLUSTRATIONS

Peter Bull 12-13, 16, 18, 20, 21 (top), 22, 23, 26-27, 28-29, 32, 33, 34-35, 42-43, 46, 49, 50-51, 53, 61, 69, 71
Garden Studio Steve Noon 8-9, Nick Shewring 10-11, 14-15
Jason Lewis 17, 21 (bottom), 30, 31, 47 (top), 48, 54, 55, 59, 62, 65, 66-67, 68, 70, 72
Linden Artists Mick Loates 71, Alan Male 62-63, 64, John Rignall 36-37, 40-41, Brian Watson 56-57, 58